HARDROCK GOLD

Books by Tom Morrison

Goldmining in Western Merioneth (Llandysul, U.K., 1975)
Cornwall's Central Mines: The Northern District, 1810–1895
    (Penzance, U.K., 1980)
Cornwall's Central Mines: The Southern District, 1810–1895
    (Penzance, U.K., 1983)
The History of the Rammelsberg Mine (London, 1989), a translation
    of *Geschichte des Rammelsberger Bergbaues,* by W. Bornhardt
    (Berlin, 1931)
To Fly through the Air: The Experience of Learning to Fly (Ames,
    Iowa, 1991)
Weather for the New Pilot (Ames, Iowa, 1991)
Hardrock Gold: A Miner's Tale (Norman, 1992)

HARDROCK GOLD
A MINER'S TALE

By Tom Morrison

With illustrations by
Cherry Hunter

University of Oklahoma Press
Norman and London

For Paul and Brenda

Library of Congress Cataloging-in-Publication Data

Morrison, Tom (Tom A.)
    Hardrock gold : a miner's tale / by Tom Morrison : with
illustrations by Cherry Hunter. — 1st ed.
        p.   cm.
    Includes index.
    ISBN 0-8061-2442-3 (alk. paper)
        1. Gold mines and mining—Canada—History—20th century.
    2. Morrison, Tom (Tom A.)   3. Gold miners—Canada—Biography.
    I. Title.
    TN424.C3M66   1992
622'.092—dc20
[B]   92-54130
                                                              CIP

The paper in this book meets the guidelines for permanence and du-
rability of the Committee on Production Guidelines for Book Longev-
ity of the Council on Library Resources, Inc. ∞

Contents

ACKNOWLEDGMENTS

To Sandy Livingstone-Learmonth, Ted Rowe, Edgar Robbins, Tim Grose, Albert Rowe, Ron Earl, Gerd Drohne, Bob Drynan, Vic Burchat, Bill Miller, Bob Moylan, Tony Horbul, Eddie Richardson, Ed Hunter, Red McKinnon, and many other stout-hearted, upright men, the salt of the earth, who passed on their wisdom, their lore, and their philosophy, I am endlessly grateful.

TOM MORRISON

Vancouver, British Columbia

HARDROCK GOLD

Now it is a strange thing, but things that are good to have and days that are good to spend are soon told about, and not much to listen to; while things that are uncomfortable, palpitating, and even gruesome, may make a good tale, and take a deal of telling anyway.—Tolkien.

We act as though comfort and luxury were the chief requirements of life, when all we need to make us really happy is something to be enthusiastic about.—Kingsley.

PROLOGUE

Think of darkness—not mere night darkness but thick blackness unrelieved by the faintest vestige of light. Think, too, of silence, total and complete. This is the true darkness, the true silence, older than time. You hear your watch ticking, your heart beating, your blood running in your ears. A sound is printed on the silence, once—the plink of a drip of water falling into a shallow puddle. Its small echo is swallowed by the silence and the darkness.

Turn your light on. That's right, turn your light on. Reach up with your right hand and twist the knurled knob on the mine lamp, which rests in a clip on your hard hat. You can now see your surroundings in a dim, yellow light. You are in a tunnel, its craggy walls and roof composed of gray rock. The roof is about nine feet high; the walls are about nine feet apart. Along the floor runs a narrow-gauge railroad. You can see only the rails above the floor of mud and compacted rubble. The wet floor gleams under the light of your lamp; puddles reflect its beam, which wanders on the roof. Shoulder-high along one wall runs a bundle of rusty pipes and cables hung by chains from eyebolts driven into the rock. In each direction the tunnel runs away into the all-encompassing darkness. You are in a mine.

You could be in almost any hardrock mine anywhere in the world. ("Hardrock mining" is practically synonymous with the mining of metals; we will encounter this term frequently.) It so happens that you are half a mile underground in a Canadian gold mine. Gold, you say? Gold, the maker of fortunes? Gold, the ultimate denominator of wealth? Gold, which is said to drive men mad? Here? Is it for gold that men have hewn this tunnel by brute force from the living rock? Well, yes, it is—but steady on, my friend. When the ore comes out of the ground, it is only gray rock. It takes four tons, sometimes as little as three, to make an ounce of gold. This here, though, is just rock. In some places it contains gold, in most places not. You see the odd speck from time to time if you know where to look, but not here. This is just a tunnel driven to where the gold is—a "drift" they call it.

Our footfalls echo in the darkness that recedes ahead of us and fills in behind us. We become aware of a faint rattling coming to us through the rock, the sound of an iron bar tapping on a cracked tile, but repeated two thousand times a minute. It goes on for two minutes or so, stops, begins again tentatively, then continues. It is the sound of someone drilling blasting holes that will be filled tight with explosives, perhaps later tonight, and blasted. An instant of rending rock riven with flame, dust, smoke, terrible concussion, and flying rock, instantly begun, instantly completed, leaving a pile of hot and stinking rubble, maybe fifty, maybe a hundred tons, just faintly stained, faintly contaminated with a chemical impurity—gold.

We come to the top of a wooden ladder projecting through a hole that leads down from the side of our tunnel. Trusting its apparently frail structure, we clamber down into the darkness. The passageway through the rock that surrounds us is steep. After climbing down a hundred feet or so, we come through the roof of a chamber perhaps ten feet high and thirty feet wide that wanders off into the darkness. The air of the stope carries the earthy smell of freshly blasted rock and the faint, acrid reek of blasting fumes. The sound that came to us faintly as we climbed down the ladders now thunders at us, hard, deep, and solid.

We turn a corner of the rambling cavern to discover the source of this pandemonium, a miner drilling with a pneumatic rockdrill, alone in the darkness of the mine. He is dressed in thick clothing covered by a suit of oilskins, once yellow but now well worn and impregnated with gray mud and black grease. On his head is a hard hat like those we are wearing, carrying an electric lamp. His lamp illuminates his work, bringing small things into sharp focus, such as the oil and grit on the back of his gloved hand, while all around him is darkness and fog, things half seen and lurking dangers, some of them lethal but innocently concealed. His shrewd recognition of these things, his vigilance and cunning, keep him alive and unhurt.

Why is he here? He does not have to be here. No one does.

Here? At night? Half a mile down a hole in northern Canada? Don't be absurd! He could be in bed and asleep and get up tomorrow to drive a truck or work on a construction site. He could be a welder or an auto mechanic, or he could work in a plant making washing machines. He does not have to be a miner working the night shift in a Canadian gold mine.

He is here partly because he is well paid. But could money alone be sufficient compensation for the endless hard work, dirt, darkness, and danger? No, there must be some other reason. This man is here because at heart he is a miner. He can look anybody in the eye and say so with pride and more than a touch of arrogance, for this is a tough industry, fiercely chauvinistic, and bristling with machismo. Thousands of men have worked in mines without ever having been miners. Those who are miners reserve their use of that term to describe not only a form of employment and a measure of skill acquired through hard experience but also a collection of attitudes and traits of character that will become apparent to us as we follow this tale.

As a wise and good man called Ted Rowe once said, mining is not just a job; it is a way of life, a vocation. You must love it, or you must hate it; there is seldom a middle ground. If you hate it, you have no place in it. If you love it, no other way of life is as satisfying, and you must follow it wherever it may lead you. With good reason Georgius Agricola described it four centuries ago as "a calling of peculiar dignity."

There is a breed, a species of this genus, known as a tramp miner, or a packsack miner. The tramp miner is a wanderer, a free spirit, an adventurer—indeed an entrepreneur in a special sort of way. "Nothing ventured, nothing gained" could serve as his motto. Personnel managers and the corporate hierarchy are his bane, and he is theirs.

"Why did you quit?" asked the hometown miner, proud of his ten years with the company. The tramp miner smiled, shrugged his shoulders, and replied, "Why not?" Between them lay a gulf of mutual incomprehension.

To some people, not least its owners, the purpose of a mine or tunnel is a matter of vital concern. To what end is it dug with so much labor and expense? To the miner, this consideration is of surprisingly small concern. His attitude is that if someone wants a hole in the ground, he will be happy to supply one.

This book is a tale of the miner's underground world in general, and of gold mines in particular. It is told, however, to you,

who may never have been in a mine. If you read tales of tall
ships at sea, or even of flight to the moon, you can at least imagine what the author is telling you. But it is difficult indeed to
conjure up a mental picture of that subterranean world of caverns and passages that some regard as the very essence of se-
crecy and terror.

Nearly everything about mining is described in a vigorous
but obscure technical jargon. My favorite, from a handbook well
known in the industry, runs: "The back is then blasted down to
fill the pocket and, after the pockets are hogged together, a lev-
elling-up cut is taken out of the back. A finger raise is driven
from the end of the scram up into the stope and a bearing set
for the cribbed raise is constructed." The author adds the sage
warning: "Care must be taken to air the pocket out after each
hog round is blasted." Such prose may cause a serious communication problem! I will do my best to clarify what has to be
explained without becoming entangled in explanation for its
own sake.

I must ask the pardon of professional mining people for a
certain necessary simplification and for using terms that are
more explicit to the layperson than the more usual mining
terms—"roof" instead of "back," for example. In general, the
terminology is that current in northern Ontario, which hosts
one of the biggest hardrock mining industries in the world.
Terms differ from mining district to mining district, and there
is no merit in confusing the reader with a multiplicity of terms.
Local terminology is used either when describing something
that has no equivalent elsewhere or when there is some particular reason for its use. If the book contains errors or misconcep-
tions, it just shows that no one knows everything.

I have not shied away from passing on tales and legends and
rumors, although I have been careful to mark them as such. We
live, far more than we sometimes realize, in a world of hearsay,
accepted facts, and what we have reason to suppose is true.
These rumors, legends, and tales half heard are a part of the
miner's life just as much as the mine in which he spends so
many hours isolated from all but a few of his fellow human
beings. I have tried to sift the probable from the improbable and
have certainly been at pains to avoid making damaging or irresponsible statements. I have also tried to be fair to both the miners and the mining companies.

This book consists of tales and episodes, some of them surprising, bizarre, even macabre. Like the scene in the stope, some are lit in sharp focus, while their surroundings remain blurred and dark. It is not a biography or a technical or historical treatise. It is a tale of mines and mining camps and the people in them. (We will often encounter this term "mining camp." It refers to a mining district—which may contain one or more solidly built towns—rather than to the collection of tents and shacks from which the term sprang.)

Not until I lived in a city did I realize how alien the mine and the mining camp are to city dwellers, who form, after all, the majority of the population of North America. Every event that I recount, however strange, is true (errors, omissions, and tricks of memory excepted). This book is written without malice. In most cases I have used fictitious personal names, and I have concealed several corporate identities.

But enough of this preamble! On with the story! Come with me to a peninsula jutting out from southwestern England into the wild Atlantic, a windswept land of low, green hills, sweeping moors of granite and heather, and little slate-roofed towns—a land called Cornwall.

TIN-MINING COUNTRY

The morning of Monday, May 8, 1972, dawned cool and bright in West Cornwall. Rain had fallen during the night, but now the warmth of the sun burned the rags of mist from the granite moorlands. Small fair-weather clouds rode the stiff breeze from the sea, casting their flying shadows on the tin-mining country around Camborne and Redruth.

A lane climbed obliquely up the side of a shallow valley from the village of Tuckingmill to a low bluff, where it divided into two. One branch left the valley altogether to wander among the ruins and rock dumps that had once been Dolcoath mine. The other dipped down to the valley bottom again, burrowed under a railroad embankment, and disappeared to the south in a cleft between the hills. The valley itself was a wasteland of brambles, tussocky grass, rubble, and building foundations of masonry and concrete. This same disorder spread away to the east along the foot of Carn Brea hill. In the valley bottom a tethered goat grazed in a small, green field. In the middle of the valley a stream frothed and burbled, red with mine waste, between banks thickly overgrown and encumbered with scrap iron.

From this vantage point the lane afforded a view toward South Crofty mine, about half a mile distant. The view was certainly neither striking nor beautiful but, for better or worse, it is the starting point of this story.

The quiet of the morning was broken by the whine of a Morris 1000 making its way up the lane in low gear. It reached the crest, where the lane divided and turned off onto a patch of gravel. The driver got out and looked about him, filling his lungs with the salt wind blowing wild and strong from the sea. He looked across the valley at the angular structures of the mine and listened to its

stony rumbling, brought to him faintly on the wind. He was a young man of twenty-two, lithe and active, of the type turned out by the English public schools described unkindly, but not unfairly, as "ready for anything and good for nothing"—well educated but with little experience of the world and a hazy conception of his place in it.

As Cortez in legend stood silent upon a peak in Darien in rapt contemplation of the vast Pacific, so too did the young man now stand gazing at his future. Like Cortez, his path to this place had not been trouble-free, and in some respects the onward journey would be as stony as the ground now beneath his feet.

His first experience of underground mining had been a visit to the caverns of a slate mine in Wales in 1966. The rock called to him as some men are called by the sea. Paradoxically, it was a time of personal turmoil and failure that had given him both the freedom and the need to answer that call and seek his living in the depths of the mines.

Could the young man's gaze have penetrated time as well as space, he would have been astonished at what the next fifteen years held in store. They would bring adventure in the richest abundance beyond far horizons in some of the wild places of the world. It was perhaps as well that he could not foresee the stupefying heat, vicious cold, lethal dangers, and crushing fatigue to which those adventures would expose him. Such things are best dealt with when they occur, for the path of the adventurer is not easy, and the only adventures experienced in comfort are those enjoyed vicariously from an armchair. The true adventurer will, sooner or later, willy-nilly, run his finger along the fine edge of hazard and feel the sharpness of it. He does not seek out these things deliberately, but neither does he flinch when they come his way. They are but the price that he must pay to live his life richly and to the full, for only in so doing can he find true happiness.

The young man was an adventurer—tough, cunning, and resolute, even though he was but dimly aware of those attributes and his life hitherto had required little in the way of their expression. It was with this inner strength that he regarded the mine.

No one knows when the first tin was mined in Cornwall. Certainly it was very far back in time; possibly it was in Cornwall that humankind first discovered what tin was. A dense, soft metal, easily smelted and easily worked, tin does not tarnish and imparts desirable properties to lead and copper when alloyed with them. It is not surprising that tin was regarded as a metal of great value. Its occurrence was not common but was restricted to certain localities, one of which was Cornwall.

Like Brittany, Wales, Ireland, and Scotland, Cornwall had been one of the last refuges of the Celts when they were displaced from the softer and more fertile parts of England and France by tribes of Germanic and Norse origin. Short, stocky, and black-haired, the Celts had consolidated and hardened into separate peoples of legendary toughness and resourcefulness in which spirituality and practicality were mingled in generous proportions. In Cornwall, above all, they were miners.

The harvest of the rock, the meager harvest of the thin, acid soil, and the harvest of the sea were all that the inhabitants of this windswept land had to support them. Greatest of these resources was the harvest of the rock, which was nowhere far from the surface and was plentifully exposed in sea cliffs and stony moors. Tin, and many other metals besides, occurred in veins, known as lodes, which dipped steeply down into the rock. Tin was at first mined from river gravels and then from trenches dug along the outcrops of the lodes, but many centuries ago the miners began to delve the strange world that is underground mining.

No single people has left its mark on any international industry to the same extent that the Cornish have influenced mining. In the 1500s and 1600s the preeminent masters of the art were the Germans. By that time a great industry had grown up in the forested hills of the Harz and the Erzgebirge, and farther east in Bohemia, Moravia, Hungary, Transylvania, and Poland. This industry was innocent of steam power or explosives. Its chief source of power was falling water. The structural material for its many ingenious contrivances was wood. For the most part the rock was carved out with the hammer and chisel, which are the international symbols of mining to this day.

If the Germans did not teach mining to the Cornish, they certainly helped them to teach themselves. By the 1700s, however, the German mines had reached their apogee. With the invention of steam pumping and hoisting engines in England, it was the Cornish who next emerged supreme.

As long ago as the 1580s an observant traveler through Cornwall remarked that Camborne was "a churche standinge among the barrayne hills," and Redruth "a hamlet annexed to uni-re-druth, where are manie tynn workes, both Stream and lode workes." The miners were at work underground even in those days, and the same observer commented that "in the Loade workes they worke all by lighte of Candle, and therfore harde to follow the Loade; which goeth sometimes sloping, sometimes directe, and oftentimes they loose it, and are much trowbled to finde it againe," a statement whose essential truth remains undiminished to this day.

In the 1600s and 1700s partnerships of a few men sank pits and shafts, drove drainage tunnels (which they called adits) from the shallow valleys, and chased the copper lodes and their rich ore, leaving holes in the ground and piles of earth and rocks like badgers. The mining leases granted to them by the lords of the manor were therefore called setts, an English term for a badger's den. They dug dozens of damp holes a hundred or two hundred feet deep and gave their mines names like Copper Tankard, Wheal Plosh, Penhellick, Wheal Crofty, and Dolcoath. One was owned by a man named Cook, who replied, whenever asked, that the lode was as wide as his kitchen, so the mine was known as Cook's Kitchen. When the miners plumbed their shafts with a rock on a piece of twine, they measured the depth, like sailors at sea, in fathoms.

The mines were wet, as rainwater trickled into the workings remorselessly. As the miners followed the lodes lengthways and downward, the problem grew worse. Not even the utmost exertions of horses and men, pumping and bailing, could keep the mines drained. They might succeed in the dry months of summer and early autumn, but the winter rains came curling in on the loud winds from the sea, and water poured into the workings, drowning them as surely as the hold of a sinking ship. One by one the mines were left to flood, or the miners did what they could by fossicking about in ground drained by adits. Pumps driven by waterwheels did something to ease this state of affairs, but the flow and head of the streams running off the narrow peninsula of Cornwall limited the potential of this source of power.

Yet some mines were so rich that no effort or expense was spared to keep them drained and working. Men who had money but no direct experience of mining took a venture in such mines, and while many of them lost every penny they advanced, some reaped rewards that were rich beyond belief. As soon as the miners began to need equipment beyond that of the most basic description, capital, labor, and technical management entered the picture as three essential and, for the most part, separate components of the operation.

At these great and rich mines the adventurers put to work monstrous contraptions of iron and wood called fire engines— the invention of a cunning artificer named Thomas Newcomen— which drove pumps in the mine by means of rods running down the shaft. Sheltered in massive masonry houses, these engines consumed coal by the cartload, which had to be hauled from the harbors to the mines along winding lanes of mud and rocks. The engines wheezed and creaked and hissed, leaking hot water and wet steam. Their ponderous motion pumped rivers of water from the mines with no more human effort than that needed to shovel coal into the furnace and keep the machinery in repair.

In the late 1760s a solid hill of copper ore was found on Anglesey, an island separated from the north coast of Wales by a sea strait no wider than a fair-sized river. At Parys Mountain the ore was dug from a pit in such quantities that the price of copper was depressed, and the narrow-vein underground mines of Cornwall were forced out of business. In twenty-five years, however, the ore accessible in the Parys Mountain pit was exhausted, so this state of affairs did not last, and the Napoleonic Wars added further to the value of the metal. As mining in Cornwall recovered, some mines could be reopened by little more than lighting a fire in the furnace of the pumping engine, which would sometimes resume its task as if it had stopped work yesterday instead of standing idle for twenty years.

The Victorian era was the heyday of the Cornish mines. This was due in part to the great improvements to Mr. Newcomen's engine effected by Watt, Trevithick, and others, with the result that the engines became more efficient in their use of coal and were applied to hoisting and driving ore-processing machinery as well as to pumping water from the mines. Hundreds of engines were built, some weighing as much as 500 tons. They were erected in houses of granite masonry with circular brick furnace stacks, some of which were nearly 100 feet tall.

During the 1830s great mines, employing a thousand men each, came into being along the north front of a granite hill known as Carn Brea. As often as not, they arose from the amalgamation or reopening of workings formerly known by some other name—East Pool, Tincroft, the Carn Brea mines, Dolcoath, and East Wheal Crofty. As the miners followed their lodes down into the granite from the overlying killas, copper ore turned to tin.

The shafts were nothing but erratic passageways sunk on the lodes. Where the lode was inclined, so was the shaft. If the lode reversed its inclination, so did the shaft. Some shafts reversed their inclination twice or more, so that in a vertical cross-section they looked like a stretched Z. This made it quite a business to drag a bucketful of ore from depths of 1,000 to 1,500 feet. So insecure and so dangerous were these hoisting arrangements that the miners climbed ladders to and from their work, in spite of the immense effort which that entailed.

As the workings deepened, they became hotter and more difficult to ventilate, with the result that heat and air so foul that in places a candle would scarcely burn were added to the burden of the miner's work. In places, advancing headings tapping the heat of the fresh rock were so hot that they had to be left to cool under the influence of whatever faint air currents there might be, until the temperature in the heading had dropped to a point where men could do useful work there. Nevertheless, in spite of bitter depressions at ten-year intervals in the late 1820s, 1830s, and 1840s, the mining industry grew in size and prosperity.

Life in those days was hard and precarious by our standards. Even so, few lives were as hard and precarious as that of a miner. Before villages grew up around the mines, and sometimes afterward, a Cornish miner would often have to walk many miles to his work in all weather, sometimes to a mine remotely situated on windswept moorlands. Carrying tools and supplies with him, he would climb hundreds of feet down crooked and ill-maintained ladders and would follow tortuous passageways to his working place, the total darkness broken only by the guttering, smoky flame of his tallow candle.

He would spend his shift hammering on the butt end of a drill steel; manhandling broken rock and hauling it to the shaft by hand barrow, by wheelbarrow, or in primitive mine cars over rough trackage; timbering insecure workings; and engaging in equally laborious occupations. At the end of the shift, he would have to climb back to the surface. At best, his "going-home clothes" awaited him, warm and dry, hanging on a nail in the boiler house that supplied steam to the pumping engine. He might have a chance to wash in hot water drawn from the engine's condenser well. At worst, he would have to walk home in his mining clothes, fouled and soaked with mud, water, and sweat, to whatever he had arranged for himself in the way of a home. Injury, maiming, or sudden death, as well as slow deterioration of health, were ever-present dangers, with slender provision against the consequences.

Nevertheless, then as now, the ranks of miners included the industrious and the idle, the thrifty and the thriftless, the sober and the drunken, those who were content because the sun rose in the morning, and those who complained about anything and everything around them. People—then as now—made the best of what they had and made merry when they had the chance.

The glory days of the mines were the 1850s. With technically advanced machinery, newly built and in the peak of its condition, the mines tapped rich copper ores from relatively shallow depths at a time of high copper prices.

Mines were often owned by fewer than a hundred shareholders, who were local people and conducted their affairs on the simple, but effective, "cost book" system. In prosperous times convivial gatherings of adventurers and mine staff were held at the mine office to review accounts for the preceding two or three months, divide profits, and discuss the mine's future. The meeting would be followed by a banquet at which toasts were drunk, and if affairs were going exceptionally well, handsome rewards were given to the mine captain and his henchmen. The miners were not necessarily left out, either, and from time to time a company might arrange excursions, picnics, and other festivities.

To be sure, mining entailed hardship and danger, but in real life such things are seldom unrelieved. Too many social historians are earnest folk who neglect the basic human propensity for having fun, whether quietly or boisterously, according to temperament—and Cornish people are second to none in their love of song and their often caustic wit.

Mines by the hundred were opened in the green hills around Camborne and Redruth and all over Cornwall. Historians have suggested that, at one time or another, something like 3,000 mines have been worked in the mining districts of Cornwall and Devon, although amalgamations, subdivisions, name changes, and interconnected workings make this a difficult figure to ascertain.

Just as in any thriving mining country, some men stayed at one mine all their lives, if the mine should last that long, while others tramped around restlessly from one to another or migrated overseas. The Cornish system was conducive to tramping from one mine to another because much of the production and development work was let out as contracts of a month's duration to crews of three or four men. Only gradually did this system give way to regular employment with a base wage plus incentive bonus, which is almost universal in the present-day hardrock mining industry.

As the British Empire expanded across the world and an English-speaking North America fulfilled its manifest destiny, the demand arose for people who knew the underground world and could exploit the unheard-of mineral wealth that was discovered in North America, Africa, India, and Australia. Moreover, the call was not merely for hewers of wood and drawers of water, although the Cornish knew these things intimately, but for those whose knowledge of underground mining was so profound that they could efficiently organize the work of others. Such were the Cornish. And Cornish they remained. Scattered to the four winds, both by restlessness and by grinding poverty at home, it was for Cornwall that they longed; it was little Cornwalls that they tried to create in the wildernesses where they labored; it was the slow drumroll of the surf pounding on the coasts of Cornwall that they heard thundering in their dreams; and it was as Cornish men and women that they were laid to rest in homesick graves in the ends of the earth.

After the middle 1860s, the tide turned against the Cornish mines. Their narrow, crooked workings reaching to ever-greater depths, sometimes as much as 2,000 feet from the surface, strangled attempts at efficient production. The maintenance of their now-aging machinery became a nightmare. Some mines remained flooded in their deepest workings by rain and pump breakdowns for months on end. The tin was all too often a will-o'-the-wisp, a little here, a little there, which soaked up money and produced little or nothing in return. Tin ore needed a more elaborate treatment plant than copper ore to convert the raw product of the mine into a concentrate that could be smelted. The price of the metal was wildly unstable. Huge deposits of tin were found in river gravels in Malaya and in hardrock deposits in the Andes and Tasmania, which put the Cornish mines into the shadows.

As the Cornish mines needed more capital, they had to seek it further afield in the commercial centers of London and the industrial Midlands. The cost-book system worked well for local shareholders, but for investors from "up the line" it looked like, and sometimes was, a den of thieves. Besides, why should a man risk his money in the mare's nest of Cornish mining when a mine like Broken Hill in Australia paid more dividends in a year than all the mines of Cornwall combined?

Under the cost-book system many a Cornish mine was owned by local smelter owners who purchased its tin concentrates, and by merchants who supplied it with machinery and raw materials. In such circumstances a mine could go on for decades as a closed system in which everyone was paid and benefited from the mine's continued operation, but which yielded no net profit. The working of the system was not necessarily dishonest (although in some cases it was), but the existence of such a system gave Cornish mining a bad name among potential investors from outside.

One by one the mines closed down. By 1900 few were still active. The workings were left to flood, and parts of them collapsed in the course of time. As much of the machinery as possible was sold for use elsewhere or for scrap. The rest was abandoned where it was, to weather and rust and rot in the sunlight and the rain.

It might be thought that the most durable surface structure of a mine would be the headframe, the latticework tower over the shaft that supports sheave wheels, skip dumps, and other paraphernalia. These structures are common relics of mining all through the American West. However, those in Cornwall had mostly been built of timber, in which case they rotted in the damp air and fell or were pulled down. In rare instances they were of steel, in which case they were sold for reuse or scrapped.

The remains that survived in Cornwall were the fortresslike granite enginehouses, which almost nothing short of dynamite could destroy. These buildings, in varying stages of ruination, dotted the landscape and marked where the mines had been. Wherever you go in West Cornwall, you may see, silhouetted on an exposed ridge crest, the enginehouse of some forgotten mine.

The remains of mining can be seen all around Camborne and Redruth as spreads of marshy, moory wasteland, dotted about with enginehouses, disordered with hillocks and rock piles, and pocked with innumerable shafts. We may stand looking up at a ruined enginehouse, craning our necks so that the building itself seems to move against the sky. Even now the bob wall is streaked with black grease that dribbled down from the gudgeon pins of the engine beam, which, of course, is long gone, as is the roof of the building. Bricks are missing from the crown of the engine stack, and it looks as if more may follow. High up in the wall is a choughs' nest; its owners, disturbed by our presence, float on the wind.

At the foot of the building is a gaping shaft, partly overgrown with brambles, choked with earth, rocks, and trash, and tenuously surrounded by a few strands of barbed wire. Footpaths thread their way between the shafts and among overgrown foundations, granite or concrete plinths that were once machinery bases, and floors of buildings, which sprout grass, woods, and rusty anchor bolts in equal profusion and equal disarray. Here and there are dunes of the red sand that was discharged by the ore-treatment plants. In the bottoms of the small valleys running down from the moors are spreads of floorings and pits filled with red mud, for there the "tin-streamers" had their day, reworking the tin-bearing wastes from the mines.

In England in about 1970 it was widely supposed that no tin mines were still at work in Cornwall, yet this was not true. Prominent between Camborne and Redruth were two steel headframes, surrounded by an agglomeration of buildings of varied size, shape, and architecture, which stood above the shafts of South Crofty mine.

South Crofty had risen, phoenixlike, from the wreckage of the Victorian years to take in, not only the old workings of South Wheal Crofty, but all the surrounding ground that had at one time been mined by its more prestigious neighbors. It had survived setbacks of all kinds, so that it seemed indestructible.

The two shafts were a quarter of a mile apart. Both dated back to the days when gigantic Cornish engines pumped water from the workings. These two engines had been housed in immense buildings that still stood beside the shafts, being too massive and of too much historic and sentimental value to destroy.

In particular, the enginehouse beside Cook's shaft, built in 1921, was of monolithic reinforced concrete. As such, it was about as indestructible as Carn Brea hill. The engine that occupied that house wrecked itself in 1950, when the fifty-ton cast-iron rocking beam snapped in two. The resulting smash so terrified the engineman that he died of shock not long afterward.

Cook's enginehouse now stood roofless and open to the sky beside the blue-painted, steel-clad buildings of the mill.

The engine on Robinson's shaft still stood in its house, where it had retired from active duty in 1955, after a working life of almost exactly a hundred years. Built in 1854 at Hayle, a few miles from Camborne, it worked on mines to the west of Camborne in the second half of the nineteenth century before being removed to South Crofty in 1903. In more recent years, the timber headframes on the two shafts had been replaced with taller versions built of steel—A-shaped structures a hundred feet tall. These structures supported the sheave wheels, which spun in opposite directions almost continually as the cages and skips were raised and lowered in the shafts. The headframe of Robinson's shaft was surrounded by a cluster of masonry or concrete-block buildings with roofs of corrugated iron or corrugated asbestos cement.

Here and there in the nearby surroundings, the casual wanderer might come across a hole extending into the depths of the earth, surrounded by a concrete collar, covered by a padlocked grating, and exhaling a damp, earthy smell. The wanderer might then look at the headframes and over to where a deep hum and a plume of water vapor, half a mile distant, told of a huge fan exhausting foul, used-up air from the mine, and he would conclude that the mine workings beneath his feet were extensive. If he watched the sheave wheels atop the headframes, noticing how fast they turned and for how long they continued to do so, he would conclude, if he was at all of an inquiring mind, that whatever material was being hoisted was being raised from great depths. Unless our notional wanderer was totally lacking in imagination, he could hardly fail to be intrigued by this evidence of unseen endeavor hidden beneath the ground.

This, in further detail, was the scene that confronted John, for such was his name.

2

DOWN THE MINE

John awoke in the guest house to a day of pouring rain. Pale gray cloud hid the top of Carn Brea. The slate roofs and rows of squat granite houses were perhaps none too prepossessing, but the place attracted him in a way he could not define. He had opened one eye at six o'clock in the morning to hear the splash of falling rain and the hiss of tires passing on the wet street outside. By the time he got up, the construction workmen with whom he had swapped yarns the night before had gone to work. Mrs. Cross cooked breakfast for him, and thus fortified, he set out on foot for the mine.

As John crested the hill, coming up out of the Tuckingmill valley, there stood South Crofty ahead of him. It knew nothing of John and cared less. He had little enough to offer it, yet he knew that he must beat on its gates and demand entry, although he was both ignorant and fearful of what he might find inside. The forces that drove him on and the fears that held him back wrestled for control. John let them play, knowing full well that he could no more turn back than could the train that came out of a cutting and clattered on its way toward Redruth.

He approached Robinson's shaft and the mine offices along a road of cracked asphalt cratered with potholes. On one side was the back of a laundry, on the other a field containing some cows and a derelict truck. Beyond the laundry toward the shaft was a yard stacked high with timber. The smells of cow dung and fresh timber competed for dominance of the wet air. The ground about the headframe was cluttered with lengths of rail and pipe, timber, round-bellied mine cars, and pieces of machinery of obscure purpose. Almost everything was rusty, caked with red mud, and seemingly in the last stages of disintegration. In one corner was

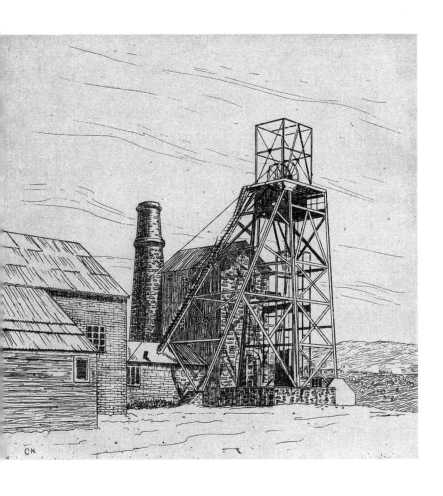

a pile of scrap. Timber was broken, splintered and frayed; steel bars were twisted like spaghetti; rails were bent and shattered; pipes were mashed and split. The whole tangle was the reddish hue of iron oxide mud. If the mine chewed up material like that, what might it not do to people?

The air quivered to the gassy thudding of a big compressor. John looked up at the towering headframe, the sheave wheels motionless for the time being, and the thick hoist cables dripping under the pouring rain. A door opened in a shack at the pit's mouth. A toothless old man, wearing torn and filthy black oil-skins and a flat-brimmed hard hat that was scarred and spattered with mud, peered out and inquired, "'Ullo, pard, what's tha want yo?" John stated his business and was directed to the personnel office. The personnel manager was at a meeting; John was to return at three o'clock that afternoon.

Uttering an inaudible malediction, John thrust his hands into the pockets of his mackintosh and slouched off into the rain. He felt apprehensive and depressed but could do nothing about it. It was still only nine o'clock; John thought that he had better try his luck at Geevor, another tin mine near the end of the Cornish peninsula, about twenty miles west of Camborne, at a place named Pendeen. He went back to the guest house, where Mrs. Cross refused to let him go any further without a cup of tea, and then drove toward Land's End.

As John followed the cliff road to Pendeen, the rain eased, and he could look down to the gray Atlantic fading into the wet mist. He had to ask directions in the village, and at length the blurred outlines of Geevor mine loomed out of the murk. The personnel manager was a woman, who said that they wanted only experienced miners. As he drove back to Camborne, John reflected that unless South Crofty came up with a job, his chances did not seem any too bright.

Just before three o'clock that afternoon, John walked once more past the shaft at South Crofty. As he did so, a bell rang a rapid series of coded signals, a human being signaling from far underground. At the rising whine of the hoist, the sheave wheels began to turn, slowly at first, then faster and faster. They continued to spin madly for several minutes, the hoist cables whipping and snaking, and then slowed to a halt as a mine cage appeared above ground. A crowd of men burst forth from it, each dressed in soaking rags, smeared from head to foot with red mud, the lamps on their hard hats still lit. As quickly as they had appeared, they vanished indoors. John stared in amazement. Even though he had been underground on occasion before, the idea struck him forcefully that these people had just come up from below the surface of the earth, where the darkness was so permanent and so complete that each man had to have a lamp with him to replace the light of the sun.

The personnel manager was a large, gruff, middle-aged man of leathery countenance who spoke with a marked foreign accent. John felt that he was face to face with an omniscient being of an all-encompassing, hard competence. He interviewed John searchingly. At last he leaned back in his chair with a trace of weariness and pointed out that, as John was obviously unaccustomed to manual labor, he might not be good enough for the mine. In that case he would be dismissed, which would ruin his chances of a future career in mining, if that was what he had in mind.

John felt like a rock-climber clawing for finger- and toeholds on a sheer granite face. With a courage he did not feel, he replied that he could only do his best and that if it was not good enough, it would not be for lack of trying. The man smiled. He wrote on a piece of paper that he folded and handed to John.

"You go see da mine ceptin. You don' tell heem nuttink you don' have to, OK?"

John's interview with the mine captain was short, and to the point. He was a small, elderly man, seated behind a steel desk in a cubbyhole of an office; he spoke in a hurried whisper.

"Eow old 'r'ee?"

"Twenty-two."

"Ev'r w'rk un'rgreound 'avee?"

"No."

"Wh'r 'ee w'rk to las', ah?"

"Unemployed."

"St'rt 'morra."

The mine store was run by a youth who had lost the forefinger and thumb of one hand. John signed for a black plastic helmet and a pair of black rubber ankle boots with steel toe caps before being passed on to the mine "dry," or changing room, where he was assigned a locker for his mine clothes and another for his street clothes, and was told to buy two padlocks. The man in charge had lost a hand. There seemed to be a number of people around the mine surface works with parts missing. John was thus processed, shuttled from place to place, and batted once more into the rain. Although apprehensive of what the morrow might bring, he was at the same time deeply grateful and happy. He was no longer unemployed and unwanted. He had been hired on.

The alarm clock woke John at five o'clock the next morning. He took the precaution of carrying sandwiches and a thermos of tea with him in an army surplus satchel that he slung over one shoulder. He did not know what the eating arrangements in the mine were, but he suspected that he had better provide for himself. He had been told to bring old clothes to work in, but he did not realize that people kept a set of filthy rags at the mine as work clothing and that they changed out of their "going-home clothes" when they clocked in. He did not make the mistake of one new man who reported for work in a suit and tie, but most people do not take long to get the message: the mine is the last graveyard of old clothes. He walked to the mine.

To the mine captain he duly reported, helmeted and booted, and was passed on to another man who assigned him a lamp and gave him a canvas belt by which to hang the lamp battery around his waist. The miner's light underground is provided by a battery-powered electric lamp. The battery case measures about $8'' \times 6'' \times 1\frac{1}{2}''$ and is slung on a belt around the waist. Four feet of rubber-sheathed cable runs from the battery to the lamp itself, which is about two and a half inches in diameter and would fit into a cupped hand. A tongue on the lamp fits into a clip on the front of the hard hat, so that the wearer automatically illuminates whatever he is looking at while his hands remain free. The battery is good for eight to twelve hours and is recharged between shifts. The whole assembly is of the most rugged construction.

John was passed on to a thickset, middle-aged man with a Somerset accent, who asked him his name and wrote it slowly in a notebook. Calling John by his first name, he led him to a miner of about John's age, whom he called Tim, and told the two of them to go into "East Pool" and continue digging a ditch. John's heart warmed to the fact that people called each other by their first names; strangers bade him good morning with a friendliness that belied their wild appearance. Another new man was standing there in the crowd, distinguished, like John, by his clean clothes and new equipment. The two of them greeted each other with wry grins. Clean clothes marked the new hand. Even the oldest of old hands sometimes appeared with some item of clean clothing or new equipment. What marks the new man, however, is that everything is new and spotless. This condition does not last.

The old hands had their own garb, each in some way unique. Only new rubber boots shine. Theirs were worn and scraped to a uniform reddish matt black and laced with blasting wire. Baggy trousers were patched, torn, and dyed by the red mud of the mine. Lamp batteries hung from frayed belts. Ragged jackets and oilskins had patches burned away by acid leaking from the lamp battery. Hard hats were painted all the colors of the rainbow. One group vied with each other to see whose hat could carry the greatest load of compacted oil and dirt. These were the "machinemen." To be a machineman was to be Somebody.

John and Tim stood in a crowded passageway outside the lamproom, waiting for the cage that would take them underground. Banter, snatches of song, and fragments of conversation swirled about them. For the most part the words were those of an intricate technical slang that has developed among generations of miners out of the need to describe things that exist only in the world of the underground. There were accents not only from Cornwall but also from other parts of England and from Poland, Italy, and perhaps other countries as well, because the mine had drawn to itself an extraordinary diversity of humankind. It would be no surprise to hear one group discussing the revelries of the night before, another the weekend cricket, and another the work of an obscure author or poet. A man would be foolish indeed to proclaim himself an expert on any subject under the sun, because someone in the crowd would know as much about that subject as he, and probably more.

Every so often, a man put his head around a door at the end of the passage and yelled a number. "Three-thirty-five!" "Three-ten!" "Two-ninety!" Each time, a cageload of miners shuffled outside with their lunch bags, water bottles, oil cans, and small tools slung about them. The crowd grew smaller.

At the cry of "Two-sixty!" Tim motioned to John with his head, and they went out into the damp morning.

The cage was just that, a steel box of girders and punched plate, in some places worn smooth, in others scaly with rust, a few flakes of paint still adhering to it. It had two decks. The top deck was reached by a flight of steps up to a landing. Tim and John crammed themselves into the bottom deck. The floor measured five feet by three, yet seven or eight men squeezed in. The bottom deck was made for short men; John had to bend his head forward; his new plastic helmet was jammed against the floor of the deck above. Feet scuffled overhead, and cold slime dripped down his neck. The cage tender rammed a steel grid into its latches, and they were ready. John heard a series of bell signals, and the cage sank into the darkness. Not a word passed among its occupants.

In years to come John would ride many different cages thousands of times into the bedrock of several different lands. Almost always, that instant of submersion in the ground, that drowning in the darkness, was greeted with silence. Conversation would cease, voices would tail off in midsentence as each man suddenly became aware of his helpless dependence on a steel wire cable, a machine, and a hoistman—and of the thousands of feet of open shaft beneath him. It is, too, a forethought of that final burial that awaits us all, that renders us all equal and of so very little account, and the speech is dashed from our lips.

The cage sped into the depths with its silent human cargo. The silence was broken only by the rustling and plopping of the guide shoes on the timber guide rails. Even though he could see almost nothing but a piece of punched steel plate a few inches from his nose, John was aware that from time to time the shaft ran through caverns in the rock. At last the cage stopped at one of these, dancing lightly on the stretch of 1,800 feet of cable. The cry "Two-sixty!" indicated the 260-fathom level. The level on which they got out was at 260 fathoms (or 1,560 feet), not below the surface, but below a drainage adit that rambles through the mines north of Carn Brea, which is itself some 30 fathoms, or 180 feet, from the surface.

John was glad to step onto firm ground again and followed the crowd along a tunnel. The only light was from the lamp that each miner wore clipped to his hard hat. Half a dozen pools of light skittered along the floor or grazed the rock walls and roof. The tunnel measured only some six and a half feet in height and width, and even that was partly encumbered by pipes. A narrow-gauge railroad track ran along the floor.

As they went farther from the shaft, the air grew warmer, and the earthy smell of the mine filled their nostrils. To left and right, abandoned passageways had been blanked off with timber and plastic sheeting, each marked with a sign: "No road." They passed through a patch of fog. Looking up, John saw the roof glittering; he thought it must be tin ore but later realized that it was only droplets of water. When he did see ore, he did not know it for what it was. It is easy enough to distinguish the greenish-black lode rock from the pepper-and-salt granite, but only a seasoned eye can tell whether the lode carries tin.

Tim turned aside into a short chamber about fifteen feet high, crudely fitted up as a toolroom. The other men went their separate ways and vanished into the maze of splaying passages. Everything of metal was corroded; streamers of white fungus hung from timbers. Tim stripped to the waist and hung his shirt on a nail; John did the same, for the air was almost oppressively warm. Tim climbed the steeply sloping rock wall using small footholds, reached behind a piece of timber, and produced two sharp picks.

"That bastard always hides the good stuff—worse than a packrat! But he thinks I don't know, and he hasn't got it figured out yet who takes it," said Tim with a grin. John laughed. Mining could evidently have its lighter side. As time went on, he found that "the lighter side" was like the jam in a jam sandwich, often what made it all worthwhile. Indeed when people started taking things too seriously and lost their sense of humor—and their sense of proportion with it—it was a bad sign.

"We'll take our kraus with us," said Tim.

"Kraus?" queried John.

Tim grinned at him. "Miners' slang. C-r-o-u-s-t, croust, lunch, food, nosh."

Tim gave one pick to John, collected a shovel and his croust bag, and off they went. John mentioned the word "spade." Tim corrected him. "In a mine it's always called a shovel or a banjo."

It seemed that they walked for a great distance through a labyrinth of deserted, red-stained tunnels with pipes along one wall and railroad tracks along the floor. They passed a side tunnel; in it was one of the miniature bulldozers that John had seen on the surface.

"See that?" said Tim. "That's an Eimco. You can get hurt pretty badly with one of those. Anything you see that you don't understand, don't touch it."

John inspected it briefly. It sat there, a machine of steel with handles and levers, four wheels, a digging bucket, pipes and rubber hoses, all oil and rusty metal. Up there on the surface no one ever saw a machine like that except its makers, because the function for which it had been built did not exist in that other world under the sky. It was specific to the world of the mine. "You can get hurt pretty badly with one of those"—it had an aura of power and violence. And it had a name, ugly, short, and specific: Eimco. "See that? That's an Eimco."

They began to walk into noise. Part of it turned into the shrilling scream of a booster fan pushing fresh air through a bellying line of flexible ducting. Once they were past that, a musical rumble reverberated down the tunnel. "Machines," said Tim, as if that one word explained everything.

A shallow ditch ran all the way beside the track. Where the ditch ended, they put down their tools and croust bags. They went into a side tunnel where there was a string of empty mine cars, took the end car, and pushed it, rumbling and bumping over the uneven track, to the end of the ditch.

They took turns breaking out the ditch with a pick and then shoveling the muck into the mine car. Tim showed John how to use the pick with short, hard strokes and no wasted effort. John was anxious, indeed desperate, to give a good account of himself and worked at a furious pace. His muscles ached; sweat poured off him in the heat, which pressed in on him, so that he could never draw enough of the hot, damp air into his lungs. He hated the hard hat, which slipped all over his head and grabbed at his hair. The plasticized-cloth ventilation ducting was covered with wet grit. The ditch had to be underneath it; it scraped his back as he worked. He wondered how long he could take it.

After a couple of hours the foreman with the Somerset accent appeared. As John stood upright, he felt ready to faint. The foreman told them to take a seat. Tim leaned the shovel against the wall and sat on it. John found a rock ledge. The foreman leaned on the mine car. He turned out to be an amiable character, very different from the hard-swearing, hard-driving mine boss that John had expected. He pointed out to John that people who fainted from working too hard too soon in the heat were of no use to anyone, and that he had better take it easier until he had gained his strength. He told them to fill their car, eat croust, dump the car, and then collect some scrap and bring it back to the shaft,

"That's Edgar," remarked Tim as the foreman went on his way. "He's the shiftboss on this level."

"What's his last name?" asked John.

"Robbins, but don't call him Mr. Robbins. He doesn't like it."

It was soon croust time. They sat down to eat. John was so tired that he could not bring himself to eat more than half a sandwich, but he drained his thermos to the dregs. Although Tim looked like a seasoned miner, he had been in the mine only two weeks since leaving the navy. He asked John a few questions about his past, which he answered evasively.

In the mine, however, no one cared where a man came from; they cared only what he was. There were men freshly out of the armed forces and men freshly out of prison; Italians who had come to England as prisoners of war, and Poles who had come as soldiers or refugees; men who had lost their jobs, and men who had never had jobs to lose. A few wandering Australians and Rhodesians passed through, eyes permanently squinting from the sun of their homelands, adding fresh and lurid expressions to those already current. There were men who had been miners and tunnelers all their lives, and their fathers before them. There were dropouts from universities and mining colleges, and those who were graduate mining engineers, but who either could not drag themselves away from Cornwall or were quietly learning some real mining. And they were all indistinguishable from one another.

Tim and John pushed the loaded car through the labyrinthine workings until they came to an open hole beside the track, seven or eight feet across and of unfathomable depth. The body of the car was seated on the car frame at each end so that it could be overturned sideways to dump its load into such a hole. A ton of mud and rock hurtled into the depths of the ore pass with a fading roar. John thought to ask what happened if you fell down an ore pass, but he dismissed the question as utterly stupid.

The rest of the shift passed easily. Returning to the toolroom to collect their shirts, they found three men sitting on planks and upturned boxes. One man held the stage, sitting on a block of wood, stark naked except for a hard hat and socks, smoking a cigar. He was in the middle of a remarkably obscene dissertation, to which John listened in astonishment. The three men ignored the newcomers, who collected their shirts and made for the shaft. "That's Albert," said Tim, as if there could be no possible doubt as to which of the three was meant.

The miners gathered at the shaft station, leaning on mine cars and locomotives. The cage came, and they crammed themselves into it. The hoist cable whisked the packed mass of tired, wet, dirty humanity up toward the fresh air and the open sky.

They dumped their lamps on the lamproom counter. Tim knocked on the window of the foremen's cubbyhole of an office to let Edgar know that they were out of the mine, and they joined the shouting, singing, whistling crowd in the steaming showers.

As John trudged back to his lodgings in the warm, sunny afternoon, he felt as if he had been through a weight-lifting course and a steam bath all rolled into one. He wore for the first time, and by no means the last, the look of the miner coming off his shift, damp hair slicked down, weary eyes squinting at the newfound day.

3

THE TIMBERMAN'S MATE

Not long after dawn the alarm clock dragged John from an exhausted sleep that had been troubled by confused dreams of a maze of rusty red tunnels and the foul obscenities of the miners. He had heard the English language used in ways he had never before thought possible. The house was silent. The jovial Mrs. Cross did not offer permanent accommodation, and John had found lodgings with Ray and Lorna Barnes on a quiet street from which he now set out on foot for the mine.

The mine looked less attractive when he was walking toward it than when it was behind him. The myriad ways in which it could dispose of people paraded before his mind's eye so forcibly that he stopped in his tracks in the valley bottom, alone, looking up at the headframe in the early morning light. With an effort he forced himself to go on. As soon as he stepped into the friendly atmosphere of the dry, he was glad that he had done so. Tim yelled a greeting to him down the line of lockers, and the day was launched.

If a man reported for work on his second shift and then on the first day of his second week, it was taken as a sign that he meant to stick with it. At about that time Edgar inquired, "John, m'son, do 'ee loike moinun'?" John replied that he did.

Tim and John went on digging their ditch, but after about a week Tim went "on machines" as Joe Salamone's "mate," and Edgar told John, "You go with Albert." Albert was the timberman on the 260-fathom level. He discovered that John worked hard, learned quickly, and could be relied on to come to work each

morning. John found Albert to be a patient teacher and full of cynical wisdom about mining and the ways of miners. Albert had been a timberman for years and was considered one of the best in the mine.

At South Crofty, at that time, work was in progress on six levels identified by their depths in fathoms below adit: 260, 290, 310, 335, 360, and 380. There were shallower levels—245, 225, 205, and so on up to the surface—but they were mostly mined out, and work on them had ceased. Access to these workings was by way of two vertical shafts. Robinson's was used for hoisting and lowering men and materials; Cook's was equipped for hoisting rock.

The tin lodes ran more or less parallel to one another, although wandering and branching, in a northeast-southwesterly direction. The tunnels that followed them on the levels were called drives ("drifts" in North America). A tunnel driven across the direction of the lodes was called a crosscut. A long crosscut from each shaft on each level gave access to most of the lodes worked on that level, while other, shorter crosscuts cut across the lodes in various places elsewhere. The drifts and crosscuts on each level were laid out on something of a rough grid pattern. The levels were connected to each other additionally by raises, or steeply inclined passageways, which could be used for access, for ventilation, as ore passes, or as some combination of those functions. The drives and raises on each lode served to identify parts of the lode worth mining. The place where a lode was mined was called a stope. All the other workings and equipment in the mine served—when all was said and done—only to give access to and remove ore from the stopes.

Most of the work was done on the day shift; some was done on the night shift, consisting mostly of "mucking out" the broken rock from the advancing drifts, crosscuts, and raises that had been drilled and blasted during the day. Men stayed on one shift or the other for years on end.

The affairs of each level were run by a shiftboss, who had a crew consisting of "machinemen," who did the drilling and blasting; "trammers," who hauled ore from the stopes to the ore passes; "O.C. men," who were general-purpose laborers; and "timbermen." (O.C. stood for "Off Contract," as distinct from the machinemen, who were paid under incentive bonus contracts and were known as contractors.) The composition of a crew changed as men were hired or left the mine, and as working places were opened up or closed down.

The timbermen were the construction workmen. They installed ladderways, built catwalks and bulkheads, installed timber arches where needed to support the roof, and built the chutes through which trammers drew ore from the stopes into mine cars. The timbermen were also the pipefitters and track repairmen. If old workings had to be fixed up, the timbermen went in there first. If any piece of gossip floated around the level, the timbermen heard it first. They cruised around places where they were supposed to be and places where they were not supposed to be. When there was work to be done, they worked hard and unstintingly. If matters happened to be less pressing, they crept away to some warm abandoned heading for an hour or two and yarned and dozed their time away. Edgar knew it, and so did Leslie Matthews, the mine captain, but they were wise enough to know that quantity and quality of work done was more important than the exact amount of time spent doing it.

John worked hard and laughed and joked because he enjoyed this strange world that he had discovered. As a consequence, he was soon accepted and identified as "Albert's mate." Nothing stays new for long underground, and he was soon indistinguishable from the other eccentric individualists that the mine had gathered to itself. There was much to learn—of chute pegs and U-irons and GWRs, of stull pieces and six-inch flats, of dog spikes and Jim Crows, furrels and Victaulic pipe, how to break a ninety-pound rail cleanly with dynamite, clay, and rockdrill oil, and how to build chutes.

Building chutes is among the finer points of the timberman's art in any mine, and South Crofty was no exception. Albert and John would collect their timber from the shaft station, where the cagemen had delivered it, and load it onto a "trolley," which was a mine car frame from which the body had been removed. They would push load after load to where the chute was to be built, perhaps half a mile "inside." They pushed the trolley around the tight corners of the tortuous drifts and crosscuts, over rickety switches and loose track joints, cursing when it derailed and the timber fell off into the ditch.

If they managed to deliver all necessary timber to the site of a new chute before starting work on it, and if everything went well, they could expect to build a chute in a day of nonstop hard work. John would get up at five o'clock in the morning and walk the two miles to the mine. Albert and he would work right through the shift without stopping for food or rest, throwing heavy timbers around in the sweltering heat, and then would hurry back to the shaft to catch the cage.

John would shower and change, walk back to his lodgings— unless Herbie overtook him and gave him a ride on the back of his motorbike—and soak in a hot bath before gorging himself on the supper that Mrs. Barnes prepared for him. He would then, perhaps, head for the Plough for a few pints of mild, weak ale.

But the job was done, and the results showed up in their pay packets as incentive bonus. The next day they would fix a leaking pipe or a piece of bad track, go to inspect the site of their next job, fetch some timber, gossip with the surveyors or samplers, dawdle over croust, and then go in behind the barricades and "No Road" signs to wander in the workings of long ago. South Crofty miners had been at work on the 260-fathom level since before 1900, so they had a good selection from which to choose. Once, they climbed all the way up to the 225.

They would approach the entrance to an old drift, barricaded and sheeted off with scrap ventilation ducting. With a furtive glance left and right, in they would go. They found drill steels from the days when shot holes were drilled by hand, and drill steel of the kind used in the old drifter machines, tools, old mine cars, and timbers baked and split by the dry heat.

Sometimes they found names or initials and dates written on the rock with the soot from candle flames or in the sharper black lines drawn with the flame of a carbide lamp. Some forgotten miner had left the mark "J R 1903" for posterity. They found cavernous mined-out stopes, known as gunnises. Sometimes only the far wall would be visible in the beams of their lamps; the rest was darkness. They could walk into the bottoms of some of these old stopes and look up into a slot in the rock 8 to 10 feet wide, inclined upward at seventy or eighty degrees for 100 feet, and perhaps 100 or 150 feet long, where the lode had been mined out. They did not venture far into such places because they could sometimes see slabs of granite cracked loose from the walls and just barely hanging where they were. Some types of rock are sticky; loose slabs will hang in place. Granite has no such stickiness, and when a slab loosens because of cracks and joints in the rock, it seems that a man need only shine his light on it and down it comes. Broken granite is often razor-sharp, and if anyone happens to be in the wrong place at the wrong time, the effect can be similar to being hit with a five-ton meat cleaver. So Albert and John tiptoed into these places, looked around briefly, and tiptoed out again. They had to take care not to burst out into the inhabited workings under some-one's nose as they hurried back to the world of the present to catch the cage.

The work of the machinemen was brutally hard all day and every day. They had no easy days, or very few. They saw only their own working places, from which they emerged at the end of the shift, their faces lined and white with exhaustion, con-cussed by the thunder of their machines, and with splitting headaches from handling dynamite. Every O.C. man talked about "going on machines" as if it were a ticket to paradise; they thought only of the much-increased pay. John saw things differently and was well content with his work. But the ma-chines are an essential part of mining, and sooner or later John had to make their acquaintance.

For about four hundred years the standard method of break-ing hard rock in mining and tunneling has been to drill holes in it, fill the holes with explosives, and blast the rock into rubble, which is then dug up and hauled away, or "mucked out," as the expression is. To maximize their effect, the holes must be drilled in certain patterns according to the size and shape of the excavation required and then must be primed so as to fire in a predetermined sequence. The group of holes used to advance a shaft, tunnel, raise, drift, or crosscut is known as a round. The dimensions of the holes and the equipment used to drill them vary widely, as do the means of removing broken rock, but the basic cycle of drill, blast, and muck remains essen-tially unchanged.

For three of those four centuries, each shot hole was drilled by a miner hammering on the butt end of a steel bar known as a drill steel, the other end of which had been formed into a chisel point. The sheer amount of this brutal labor expended in mines all over the world by miners wearing themselves out pounding steel is terrifying to contemplate.

In the latter half of the last century, machines powered by compressed air were invented to perform this labor mechani-cally. Decades elapsed before machines were built that were tougher than the miners they were supposed to assist. The real hardrock miners laughed at the "machine miners" and their machines that were forever breaking down, and went on pounding steel.

The early rockdrills were heavy and cumbersome to use. The machines and their mountings had to be manhandled into place and repositioned for each hole to be drilled. The driller's work was somewhat easier than with hand steel, but it remained hard work. The new machines were incredibly noisy, and less obvious but more dangerous, the faster rate of drilling churned out large quantities of rock dust, which scarred the machinemen's lungs and finally killed them. It was a particularly ghastly death, at that. In the early twentieth century this problem was solved by injecting water through a hole running down the center of the drill steel, so that the rock dust spewed out of the shot hole as a wet slurry as drilling progressed.

In the late nineteenth century dynamite replaced black powder as the chief explosive; it was typically supplied in one-inch-diameter cartridges eight inches long. Unwrapping the waxed-paper wrapper of a cartridge would reveal a whitish, mealy paste. Although dynamite had many desirable properties, safety being one of them, nitroglycerine fumes penetrated the skin of anyone handling it and caused a "dynamite headache." Dynamite had other useful properties. Albert remarked to John one day that if he was ever troubled by constipation, a small pinch of dynamite taken in a cup of tea had laxative powers that were nothing short of amazing.

So the machinemen came to the surface after each shift, punch-drunk and in vile tempers from the roar of the machines, plastered with mud and machine oil, and with splitting headaches. It is no wonder that the old-timers preferred to go on pounding steel; in some mines they continued to do so until the 1950s.

In the late 1940s a new, so-called lightweight rockdrill appeared on the scene. Instead of the complicated and unwieldy arrangement of bars, clamps, screwjacks, and cradles that supported the "drifters," or "Leyners," as the older machines were called, the new device was, in essence, a jackhammer equipped with a pneumatic, telescopic pusher leg. A valve bled air from the compressed air supply into the leg.

Leg machines are manufactured in two varieties. An "airleg," or "jackleg," has a hinged joint between the leg and the machine, while a "stoper," or "hammerdrill," is a straight-line, rigid unit. In general, jacklegs are used for drilling horizontal or slightly inclined holes, while stopers are used for drilling steeply inclined or vertical holes. The cunning miner will find ways to use either machine for either purpose, should the need arise. (The stoper actually dates back earlier than the jackleg, but we cannot pursue that kind of detail here.)

The new leg machines rapidly dominated underground mining, continued to do so for thirty years, and, to some extent, still do today. The acid test of a miner came to be, quite simply, his dexterity with a jackleg.

The jackleg driller's work is far from easy. The machine weighs up to 130 pounds. A sensitive valve controls the pusher leg, which reacts with startling rapidity and great force. The driller does not support the whole weight of the machine all the time, but running a machine all day nevertheless requires great stamina and sheer muscular strength. To the uninitiated, the machine is a mass of apparently solid steel of unbelievable weight, unpredictable, uncontrollable, vicious in the extreme, and capable of inflicting a wide variety of serious injuries. The initiated can use it to engrave his name in the rock and can, on occasion, run two at once, although he is unlikely to make a habit of it.

The driller spends his time soaked in sweat, oil, and dirty water, ankle-deep in mud, dimly able to see through the fog and oil-mist exploding from the exhaust, and battered by the ear-splitting thunder of the machine, made up of the impact of steel on steel and of steel on rock and the escape of compressed air at 80 to 100 pounds per square inch.

Most of the ore mined at South Crofty was stoped by the so-called shrinkage method. The miners drilled and blasted upward on the lode over a length of, perhaps, 100 to 150 feet, standing on the broken ore that they had blasted down. Broken rock occupies more space than solid, and therefore ore had to be drawn off through chutes at the bottom of the stope so as to maintain the necessary working space. When the miners had drilled and blasted their way up to the level above, the whole pile was drawn off, leaving an empty slot in the rock where the lode had been.

The method depended for its successful operation on an even downward movement of the whole muckpile as the trammers drew it through the chutes. Sometimes the muckpile would "hang up," and the trammers would have to push dynamite charges up into the hung-up chute on poles to blast the hang-up and make it collapse. The problem was that if a substantial void had developed inside the muckpile, the whole thing would come down at once, leaving far too much space between the broken ore and the stope face, so that the miners would have to build catwalks to reach the face.

It was essential for the miners to make sure that the muckpile was subsiding properly wherever and whenever the trammers were drawing ore. Otherwise a hang-up could develop inside the muckpile without anyone being aware of it. Should anyone be standing above a hang-up when it chose to let go, that was the end of him. John narrowly escaped this fate.

A chute had hung up, and an open chimney had formed in the muckpile above it. John happened to be there when the trammers were about to blast it down, so they sent him up into the stope to warn the miners. Unwittingly, through inexperience, he walked over the top of the chimney. The crust of broken rock on which he walked—twice—was so thin that the trammers forty feet below heard his footsteps. He returned to find them white and shaking from the expectation of having to pull his remains out through the chute. They cursed him roundly for his stupidity, although it was a month or two before he figured out why, and how narrow had been his escape.

One morning, Wayne, a newly trained machineman working for Edgar, needed a mate for the shift. John drew the short straw.

Each machineman's equipment included a bag of drill steel. The drill steels were made from seven-eighths-inch hexagonal bar stock and came in lengths of 2, 4, 6, 8, 10, and 12 feet. One end of each steel was formed into a shank to fit the chuck on the machine; the other end was made with a chisel-shaped tungsten carbide insert for drilling the rock. The inserts had to be sharpened at the end of every shift of use. Many, if not most, mines had gotten around to using bits that could be detached from the steel, so that only the bits needed to be taken to the surface for sharpening, and not the whole steels. South Crofty still used one-piece steels—because of the same conservatism that caused the miners to resist the introduction of electric hat lamps until 1966, some fifty years after they had become common in the rest of the hardrock mining industry. There were no bit sharpeners underground, so each machineman's bag of steel had to go to the surface at the end of the shift. The steels were sharpened during the night, and the bags were dumped on the levels before the start of the day shift. It is easy to guess who carried the bag of steel to and from the working place.

When Wayne and John got out of the cage on 260, Wayne merely pointed to the bag of steel lying on the shaft station floor and went his way. John hoisted the bundle onto his shoulder. It must have weighed fifty or sixty pounds. He had to stoop frequently to avoid entangling the steels in chutes, pipes, and other bandersnatches, and he spent the time in between trying to invent new and more expressive swearwords. To get the bundle into the stope, he had to drag it up fifty feet of crooked ladders. The stope was four or five feet wide, and in places the muckpile of coarse broken rock was only four feet from the roof.

John found Wayne sitting in the stope drinking tea. He directed John to bring the jackleg machine, its hoses, and the drill steels to where he wanted to drill. The hoses and their metal fittings snagged on every projecting rock throughout the length of the stope, and there were many. The low roof of the stope forced John into a crouch, as he tripped and stumbled over rocks. Projections of the rock roof gouged his back or brought him up short when he banged his head into them, provoking a grumbling discourse of muttered profanity. Wayne lit another cigarette and examined the rock in a professional manner.

When all was ready, John's job was to hold the end of the drill steel each time Wayne started a hole. For one thing, the point of the steel had to start off in the right place. For another, it had to stay there until it had broken out a niche for itself. Then John could stand aside while Wayne drilled and then would exchange each steel for a longer one as the shot holes were deepened. A skilled machineman can start a hole on any rock face in any direction with no assistance whatever. Wayne was not a skilled machineman. He was a scrawny kid and wrestled with the machine while John held grimly onto the steel.

Wayne would yell through the pervasive hiss of leaking compressed air, "There!" (vaguely flashing his light onto some point on the rock). "No, up a bit. No, over there. Down! Down!" (Abuse sotto voce, which John pretended not to hear.) "OK. There!" Scats of rock and a spray of water exploded in John's face as Wayne turned the machine on. Small rocks fell on his wrists and drew blood. The chisel bit jumped across the rock. The machine bucked and reared; Wayne fought with it and swore. And they were back to square one.

One by one they got the holes drilled. John was soaked in sweat and gritty water. The stope filled with the pealing thunder of the machine and the fog from its exhaust. Croust time came and went, and still Wayne fought with the machine. John's stomach was just beginning to rumble at full volume, when Wayne said, "That's the last hole. You go back shaft and get the dynamite."

It was a quarter of a mile back to the shaft, where Edgar was issuing dynamite from the magazine, a stub drift with a concrete wall and a steel door. The other machinemen's mates had long since collected their dynamite and carried it off into the labyrinthine workings. Nearly a hundred pounds had to be carried back to the stope in two plastic buckets. John was positive that his arms stretched by an inch. He had not bothered his bewildered mind as to how Wayne might have been occupying his time. The question answered itself. Wayne had evidently enjoyed his croust and was just smoking a postprandial cigarette.

The machine and hoses lay where they had been used. While Wayne loaded the shot holes for the blast, it was John's task to drag the gear to the far end of the stope by the same laborious procedure as at the start of the shift.

They fired the blast electrically from down in the level. The sole evidence was a dull booming and a series of soft concussions in the air. Wayne led the way back to the shaft, John stumbling along behind him with the bag of drill steel. On the shaft station, John thought that he might have time for a cup of tea before the cage came, but he was wrong. He burned his tongue trying to drink it, threw the rest onto the floor with a malediction so disgusting that even the cageman looked at him in surprise, and pushed his way onto the cage. He was not sorry to resume his work with Albert the next day.

4

THE LIGHTER SIDE

John would long remember those summer mornings when he got up to go to work at South Crofty. He awoke each morning to the sound of doves cooing among the cherry blossoms across the road in the quiet of a summer dawn. He ate his breakfast standing in the kitchen, the silence broken only by the ticking of the clock, while the light grew outside. He filled his thermos and an aluminum water bottle with half a gallon of tea and furtively unwrapped a corner of his sandwiches to see what Mrs. Barnes had put into them. He was always grateful for the little surprise, re-experienced at croust time, because they were never the same two days in a row.

Shutting the door quietly behind him, croust bag slung over his shoulder, he stepped out into the glory of the new day. His route was little used at that hour. Sometimes a man was standing on a street corner, waiting for his ride, and they greeted each other gruffly, but otherwise he made his way alone through the sleeping outskirts of Camborne, where birds sang their first tentative notes, and the scent of flowering gardens hung rich and cloying on the still air. The gently climbing road crested on a patch of open ground; there to the east were Carn Brea hill and the mines.

Soon the quiet of the early morning was no more, as a throng of miners converged on South Crofty on mopeds, motorcycles, scooters, and bicycles; in cars and vans; and on foot, hatted and hatless, in T-shirts and anoraks, jackets and old raincoats, jeans, corduroys, and parts of cast-off suits, all with the red-stained croust bag over one shoulder. They swarmed in through the doors of the dry with its sharp, warm smell, and so the shift began.

Edgar boasted of the proficiency of his two timbermen, but between them they conspired to tease him almost beyond endurance. He had once been a deputy in a Somerset coal mine. (The titles given to the ranks of mine officials in British coal mines differed from those of hardrock mining and included such titles as "deputy," "overman," and "shotfirer.") Albert nicknamed him Deputy Dog and referred to him as such, often not quite out of earshot.

A low stone wall bordered the mine yard about sixty feet from the shaft. John and some of the other O.C. men would lean on it on fine mornings before going underground and watch rabbits playing on a rock dump. The miners loitered around the headframe in the long, cool shadows as the sun climbed over the eastern hills, sat on piles of timber, or lounged on mine cars, Eimcos, locomotives, or whatever else was at hand. The crowd diminished as the cageman called the levels and cageload after cageload disappeared underground.

When the cageman yelled "Two-sixty!" Edgar's O.C. men would tread the fine line between letting him yell himself hoarse and the point where he told them to go to hell, slapped the gate onto the cage, and rang it away, leaving them to face the consequences. Edgar was often outside as well, and found this particularly exasperating. They kept a watch on the proceedings out of the corners of their eyes, while their backs expressed concentrated boredom and disinterest. The miners already on the cage watched this ritual with amusement. Finally, Edgar would explode, "DON'T YEOU OIDLE %@#$%^ &* AVEN WANT TO GO UNDERRGREOUND?" in tones that could be heard over most of the surrounding neighborhood. Then they sprinted for the cage, their muddy rags flapping and rustling about them, leaving a red-faced Edgar shaking his head.

As the South Crofty miners pushed their drifts eastward into the abandoned East Pool mine in search of unmined ore, they had to drill long pilot holes to probe for flooded workings. Dams were built in these easterly drifts in case a pilot hole should tap more water than the mine pumps could handle. The dam that was built on 260 in a sustained, round-the-clock effort was a plug of concrete ten feet thick with the door and doorframe of a ship's bulkhead built into it. When the door was finally in place, it was found to bear the inscription: "Abandon Hope All Ye What Enters These Mortal Portals." Graffiti were strictly forbidden in the mine, on pain of dismissal. Edgar knew that it was the work of his two timbermen, but Albert had supplied the text and John the calligraphy, so that each stoutly denied responsibility; Edgar was left to gnaw on timbers or express his displeasure in any other way he saw fit.

After the dam was completed, Edgar told the two of them that on no account were they to go behind it, and that part of the level became a no-man's-land, populated only by diamond drillers and other creatures of the night. Behind the dam was an antique mine car, of the type known as an old-man waggon, of which a few lay about the level, mostly upturned and used for concealing tools, unused explosives, pornographic literature, and other objects of dubious legality.

One morning, not long after he had issued this order, Edgar was walking the level and met a dilapidated old-man waggon rumbling toward him, loaded with pipe fittings and scrap timber, propelled with much sweat and blasphemy by his two timbermen. His mind flew to the old-man waggon behind the dam. He questioned them, to which Albert replied with guilty looks, half truths, and indefinite confessions, while John kept his mouth shut and wore the look of an imbecile. They helped Edgar to work himself into a towering rage, and he left them with the dire threat that they would be on the carpet in front of the underground manager after shift. The truth was that Albert and John had found another old-man waggon elsewhere on the level and had pressed it into service. When Edgar accused them of going behind the dam, the opportunity had been too good to miss. Edgar must have gone to check his evidence, however, and must have realized with chagrin that his leg had been pulled. Neither Albert nor John heard anything more about it.

Lionel Keast and his mate, Gerry, were sent to drill through a brick dam that had been found in old workings in East Pool. East Pool had been abandoned in 1945, and much information about it had been lost or forgotten. Now that South Crofty was engaged in a long-term process of extending their own workings among those of East Pool, all sorts of surprises were in store. The event was attended by the whole hierarchy of the mine staff—Gerald Pengilly, the mine manager; Nigel Ebsworth, the underground manager; Leslie Matthews, the mine captain in charge of that part of the mine; and Edgar.

The dam was unexpectedly thin, with more water pressure behind it than anyone realized. Albert and John were repairing track in the main crosscut on 260 that day. The procession that squelched past them consisted of some of the wettest people John had seen for a long time. Edgar glared at them sulfurously. Gerry, who was a light-hearted soul, brought up the rear, mimicking them all one by one. It is doubtful whether Edgar's temper was much improved by the cackling laughter that echoed along the crosscut behind him.

Most elaborate were the jokes that Albert and John wreaked on each other. Albert spent weeks implanting in John's mind the sad state of his right thumb. Allegedly, the thumb had been so seriously injured years ago that Albert had nearly lost it. Even now, it sometimes got out of joint with agonizing results. John was therefore primed and ready to render all possible aid when, one day, Albert suddenly burst out, "Pull my thumb! Pull my thumb!" with obscene blasphemy and cries of pain. John gave a good yank on the afflicted member, whereupon Albert cut loose with a fart that would have cracked window panes but that, as it was, spent its considerable energy reverberating harmlessly through the workings of the mine.

Chief among Albert's pranks was "hooking people up." Many an unsuspecting miner, leaning on a locomotive while waiting for the cage at the end of the shift, would find, when the cage arrived, that he had been tied to the locomotive with blasting wire. His companions had engaged him in conversation while the hooking-up was in progress, having themselves been victims of this prank, or knowing full well that they eventually would be. Sure enough, Albert hooked John up one day.

John bided his time. Tying one end of the blasting wire to a locomotive or other similarly immovable object was easy; tying the other end to the victim without being detected was not. However, the steel S-hooks used for hanging ventilation ducting offered an ideal solution, especially if one was purposely bent so that it would slip snugly over the victim's lamp cable. John picked up an S-hook in his travels, bent it to shape, tied a length of blasting wire to the closed loop of the hook, tripled for extra strength, put the device in his pocket, and awaited his chance.

The deed was done quickly. John absented himself from the vicinity and went to talk to someone beside the shaft. Of course, no one was going to tell Albert that he had been hooked up. The best results were achieved if the victim remained still, without making some small movement that might disclose the trick, and then tried to make a rush for the cage. So it turned out. Albert was brought up short with a wrenching jerk amid the guffawing miners. John pretended not to notice but did catch a sly glimpse of Albert's face. Albert was not the forgive-and-forget type; Albert would be looking for blood.

Albert knew the exact spot where John leaned against the wall opposite the lamproom each morning to buckle his lamp battery onto his belt. The next morning, John was so engaged when he spotted his customized S-hook lying on the floor. He finished buckling his belt and hitching it about his hips so that it was comfortable, looking about him vacantly as he did so. Albert was watching him like a hawk.

Sometimes Albert pushed his luck to the limit. During one cage ride to the surface, he was riding in the top deck of the cage. Only two or three other men were occupying the top deck, while the bottom deck, beneath the floor of punched steel plate, was fully occupied by a jammed mass of men. Albert remembered that he still had some warm tea in his thermos. Announcing to the assembled company below that he was about to urinate on them from a height, he emptied the contents of his thermos at a suitable rate of flow. The enraged screams of obscene abuse from the bottom deck had the men above laughing until the tears streamed down their faces. By the time the cage reached the surface, they had the men below calmed down enough to see the funny side. Fisticuffs were about as strictly forbidden as anything could be, so it was just as well.

The mine was not all fun and games—and besides, some men could not quite see the funny side. That was why it was not uncommon for ten new men to start work on a Monday morning, and for only three of them to report for work on the Friday. The weekend would give those three men pause for reflection, and perhaps only one or two of them would be there on the next Monday. By that time there would be another crop of newcomers.

John was fortunate in having been assigned to 260, the shallowest and coolest of the working levels, and in having gone to work for Edgar Robbins, as fine a shiftboss as any man ever worked for. If John had started out with an ill-natured partner or a sour shiftboss in the furnace heat of the deeper levels, who knows if this tale would have been written?

Apart from the dirt, darkness, and danger that are common to most mines, the main problem at South Crofty was heat, which, in places, was stupefying. As one penetrates the earth toward its center, the rock becomes warmer. The number of degrees of warming per thousand feet of depth is called the geothermal gradient and varies from one part of the world to another. Mines in the Canadian Shield, in rock that has remained undisturbed by geologic forces for a billion years, are bitterly cold in their upper levels, pleasantly warm at 3,000 feet, and hot at 6,000 feet in those mines that penetrate the earth that far. The granite of Cornwall, forged in the furnaces of the earth's mantle as little as 300 million years ago, is much warmer. Even at 2,000 feet from the surface, the heat is ferocious.

In hot mines in some parts of the world, heat is relieved by using enormous fans to force huge volumes of fresh air through generously dimensioned passageways, or even by refrigerating the air. South Crofty had no such amenities. The tangle of narrow, crooked, interconnecting workings made it a ventilation engineer's nightmare. The heat lay in wait for the miners away from the shaft, back there in the drifts and stopes and raises. Once, when John happened to be on the 360-fathom level, he stood aside to let a trainload of ore rumble by. Hot though the air already was, he could feel heat radiated by the broken ore in the train, which had come from some furnacelike stope far inside.

There was no escape from the overpowering heat. It pressed in on a man like an almost physical burden, dulling his mind, and infusing him with a terrible lethargy. In some places a man would sweat steadily even while he was sitting still. When he went to work, sweat burst from every pore. John measured the heat in a working place by the rate at which sweat dripped from the end of his nose. Most people stripped to the waist at the start of every shift; at the end of the shift their only dry clothing was the shirts and jackets they had left hanging on a timber. John spent one shift shoveling spilled rock from under a chute where the air was warm and the ventilation flow dead. After that shift, he poured sweat from his boots to splash on the floor of the dry. To balance this continual sweating, he drank a gallon to a gallon and a half of various liquids each day, almost all of which was either tea or mild English ale. Surprisingly enough, most people became acclimatized after a few weeks underground and thought nothing more of it, unless they had to work where the heat was particularly oppressive.

The ventilation flow through the mine was driven by a large fan that exhausted air through an old shaft in the west end of the mine and drew fresh, or "downcast," air down through other old shafts in the east. The eastern workings were thus supplied with fresh, cool air from the surface. By the time this air reached the western, or "upcast," end of the mine, it was turgid with heat and moisture.

The active workings on 260 were all east of Robinson's shaft. The 290, however, was active both in the east and in the west around Cook's shaft and in the section known as Tincroft. Albert and John sometimes worked for Ronnie Opie, the shiftboss on 290. Whenever Albert greeted him in the morning with the news that they would be going to Tincroft, John merely uttered two words, the second of which was "Tincroft."

Even going to Tincroft was quite a journey. After riding the cage to 290, they headed north along the main crosscut from Robinson's shaft for several hundred feet to where a drift followed a sinuous lode westward into Cook's section of the mine. The drift was baking hot and bone dry. Dust on the floor and the closeness of the walls and roof muffled the thumping of their footsteps. Their few words of conversation fell dully on the hot air. The timber of track ties and old chutes had been baked and split by the heat.

Such hot, dry, deserted workings had their own smell. It was a curious mixture of warm sand, honey, and cinnamon, given off by old timbers or by the lodes themselves. Some lodes, when cut into, were found to contain hot springs, which gushed out on the deeper levels and made a steam bath of their surroundings. One lode could be traced from level to level by its unmistakable sticky-sweet smell. Some lodes contained "vugs," or open cavities, whose walls were treasure troves of crystals. Albert showed John one vug as big as a room, containing a sticky brown substance known as mineral pitch. The mines of Cornwall were full of such curiosities as these.

As their plodding feet brought them to the intersection with Cook's north crosscut, it was like walking through the door of an oven. Heading south, they would come to where lights indicated human activity around Cook's shaft. As they passed the shaft, they might linger for a moment to watch the steel hoist cable running in the shaft like a silver snake, with a slight swaying and twitching motion. A skip would pass the level, a blur of rusty metal traveling with almost unbelievable speed and silence, hustling seven tons of rock from the loading pocket on 380 to the mill bins at the surface. Cook's south crosscut would bring them to the steamy, wet workings of Tincroft section, after they had walked a mile, traversing two and three-quarter sides of a square of rambling drifts and crosscuts.

Two-ninety Tincroft was jinxed. Anyone who does not believe in jinxes can remedy that deficiency by working in an underground mine for a few months. Albert and John went there to build a chute.

The main structural element of each chute was a pair of round logs, perhaps ten or twelve inches in diameter, and eight to ten feet long, known as stull pieces. Each one had to be lifted up and set into place at a forty-five-degree angle above the drift in the boxhole where the chute was to be. The foot of each stull was set in a hitch cut in the rock about five feet above the floor; its head butted against the opposite wall. The two timbermen had to wrestle the stulls into place with nothing more than the cunning and vigorous application of muscular strength. Generally, the job took half an hour; in the jinx-ridden Tincroft things were different.

Two stulls lay beside the track in Cook's north crosscut, but they were baked dry, split and useless. The stulls that Albert and John were told to use were sixteen inches in diameter. Conveniently, they had been left lying under a spatter of dripping water. John looked at the lengths of massive hardwood, water-logged and slimy to the grip. The miners who had driven the boxhole had driven it a little too wide, or a slab of rock had peeled off the wall, and as a result, the stulls would have to be twelve feet long. Man-handling such monsters into place was going to be a back-breaker, and if one of them got loose in the process, someone would get hurt. They rounded up some help. It took six men a whole shift to fight those two stull pieces into place.

The next day Albert and John, in turn, were roped in to help someone else. Some enthusiastic operator on the night shift had managed to somersault an Eimco, so that it lay with its wheels in the air and its bucket facing back toward the shaft. Surprisingly, the operator had escaped unscathed. When the Eimco was back on its wheels, they went and built more of their chute. The timber they had to work with was as hard as iron. John was up in the chute, kneeling on its sloping bed, trying to nail the side boards into place. Half the time, the nails just went "Bing!" when he hit them and flew off into the darkness. Albert, standing in the level below, made various sarcastic suggestions until

John invited Albert to show him how to drive a nail. Albert had no better luck, and his temper grew worse by the minute. But, of course, that had nothing to do with jinxes.

A new man was sent to help them who aspired to become a geologist or a mining engineer. At croust time Albert, John, Stan Tregwin, and a few others forgathered in a dry spot where there was a faint breath of air. Conversation turned to some of the more gruesome accidents that had befallen people in the mine.

Next day their erstwhile partner was not there. Evidently, he had concluded that this steamy darkness, populated by maniacs, beneath what he had always imagined to be solid earth, was hell with the lid off, and he was seen no more in those parts. Perhaps he fled from mining altogether and dined out for years afterward on the experiences of his brief time as a miner. Perhaps he retreated to the realm of clean clothes and air-conditioned offices, where numbers and formulae do not shout and swear and behave themselves in an unseemly manner; where risk is a financial statistic, not falling rock, blood, and broken bones; and where heat is a thermodynamic parameter, not an overpowering, inescapable enemy clawing at the mind.

Ronnie Opie passed by and inspected the half-built chute. He thought the throat a little tight for a satisfactory flow of broken ore and decided that the rock would have to be drilled and blasted. He got hold of Stan Tregwin. "Stan, m'son, 'ave us couple 'oles in 'ere, pard. Woiden the @#$% ˆ& eout li'l bit yo."

The next time Albert and John returned to make yet another piecemeal addition to what ideally would have been completed in a day, they found their handiwork blasted to rags and tatters. Only the massive stull pieces remained undisturbed. Albert and John stood there, wilting in the stupefying wet heat, surveying the wreckage, and unsuccessfully searching their minds for words adequate to express their feelings. After a long silence Albert remarked with a sigh, "Not our $%ˆ&* day, pard." There was nothing for it but to sit down and have a cup of tea before setting about cleaning up the mess.

As soon as things were back in shape, but before they had a chance to complete the chute (and collect the bonus pay that went with it), management decided that they did not need the chute after all. Ronnie Opie made sure they had something in the way of bonus in their pay packets anyway.

No matter how bad things were, how dangerous, or how frustrating, there was always something to tickle the ribald humor of the South Crofty miners. Three-eighty, the 380-fathom level at the bottom of the mine, was supposed to be the big production level. It was hotter than the hobs of hell down there, a fact that John never had to ascertain by direct experience. It was also the level where all ore from the east part of the mine was drawn through a compressed-air chute at the bottom of Robinson's ore pass and trammed across to the crusher and skip-loading station at the bottom of Cook's shaft.

Management decided to improve the track on the main line down there by replacing the wooden ties with concrete ones. With wooden ties, rails are laid down on the ties, set to gauge, and then fixed in place with track spikes hammered into the wood. Such a procedure is impossible with concrete ties, so clips must be set into the concrete at the correct gauge at the precast plant where the ties are made. Rails are laid in the clips and locked into place with metal widgets.

The ties were manufactured to the company's specifications at a local precast plant and moved underground by the ton in preparation for the big push of relaying the track. This work had to be done in the shortest possible time to avoid holding up production from Robinson's side of the mine. At that point it was discovered that the clips had been cast into the concrete at the wrong gauge. Albert had his ear to various doors and reported the resulting witch-hunt with a fiendish glee. So really, it was fun and games after all, as long as a man did not weaken.

John was happy working at South Crofty. By letting events take their course, he would eventually have become a timberman, with a spell on machines somewhere along the way. The timberman doubled as chargehand, or foreman, on his level if the shiftboss was absent. After five years or so, management would have offered him a chance to become a shiftboss. In the more distant future, he might have been promoted to be a mine captain. There he would have remained until he left the mine or retired—unless the mine closed. But John knew that he would crave for broader horizons and a deeper technical understanding of his vocation than he would ever gain by working as a miner at South Crofty. He had resolved to obtain a mining engineering degree at the Camborne School of Mines; working at South Crofty was the acid test to discover if he was cut out for mining. The answer was in the affirmative.

The sad day came, as the leaves changed color on the trees,

when he left South Crofty to take up his studies, never to return. More than once, over the years, he would look back wist-
fully and wonder if he had been wise. If he only knew it, that
unsatisfied curiosity, that decision to sacrifice known and con-
genial surroundings for the chance of uncertain gain beyond
the horizon, was the mark of the tramp miner.

5

THE THOUSAND-YEAR MINE

The train ran on across the rich farmland of northern Germany under the summer sun. John slept. A little more than a year had gone by since he worked his first shift at South Crofty. He was now about to spend the summer at the Rammelsberg mine on the northern edge of the Harz Mountains. He knew Germany, its land, language, and people. After leaving South Crofty, he wanted an underground job for the following summer, and he wanted to visit Germany again. Putting the two together resulted in his present position.

He had spent the night on a boat crossing the North Sea before boarding a train at the Hook of Holland at five o'clock in the morning for a ten-hour journey across Holland and north Germany. A wakeful night and more than twenty-four hours of continuous travel combined with the smooth motion and soporific rumble of the train to account for his comatose condition. Let us leave him to his slumber and precede him to his destination.

We become aware of a range of hills, blue in the summer haze, rising out of the plains to meet us. As we come closer, we see that they are thickly wooded, in contrast to the open farmland at their feet. Standing before them is a town whose clustered towers and spires, gleaming in the sunlight, bear witness to its ancient origins. The hills are the Harz Mountains; the town is Goslar.

The rounded hills rise 1,200 feet above the plains. One of them, indistinguishable from the rest, is known as the Rammelsberg. A little over a thousand years ago ore was found on the northwest slope of the mountain.

The Romans never reached these hills or the mineral wealth of central Europe, and therefore mining never received the impulse that it did in Mediterranean lands. In the tenth century A.D., min-

ing operations in Germany were small and scattered. They developed in size and engineering skill over the succeeding centuries until German mining engineers led the world as masters of their art. One of the world's first colleges of formal education in mining engineering was established at Freiberg, among the mines of Saxony.

The German mining industry grew up in a country divided among numerous small, self-governing states, ruled by a multitude of dukes and princes. If the mines were not owned outright by princely families, state control and regulation were often prominent. A high degree of continuity was thus assured, and many great schemes of adit drainage were undertaken when the instigators knew that their grandchildren would be the first beneficiaries. In those days, when mining was by hammer and chisel, an adit might advance six inches in a day, yet in the 1500s adits were begun that would entail miles of driving when the driving of a single mile might take a lifetime.

Overseeing this work was a hierarchy of state officials who were honored to spend their lives in the service of one princely family; in some cases the prince's mining advisor was among his most respected counsellors. The German mines, therefore, developed a tradition in which work went ahead, carefully engineered and strictly regulated, in a manner that would lead one nineteenth-century Cornish mine captain to remark contemptuously that three Cornishmen would do as much work as six Germans, and that German mining was "more theoretical than practical." Maybe there was something to be said for this attitude, for it was the Cornish who emerged as the leaders of hard-rock mining technology, mostly through the empirical design of large and—for its time—sophisticated machinery, and Cornishmen dominated the rough-and-tumble of the North American mining frontier. Many of the outstanding achievements of German mining took place in the forested hills now before us, around such towns as Clausthal, Zellerfeld, St. Andreasberg, Wildemann, Lautenthal, and at Goslar under the Rammelsberg.

The ore discovered at Rammelsberg turned out to consist of two huge lenticular masses of mixed metal sulfides—lead, zinc, copper, and iron—rich in silver, and containing gold, barium, antimony, bismuth, cadmium, mercury, nickel, cobalt, germanium, indium, thallium, gallium, and other metals. A thousand years ago the rare metals were unknown, unrecognized, and without known uses or methods of extraction; at that time the ore was valued for its silver and lead, and to a lesser extent for its copper. The total mass of ore in the two lenses was about 27 million tons, the last carload of which was mined in 1988 and was paraded through the streets of Goslar with due solemnity. In 1973, however, the mine was in full swing, and although the end was foreseen, it was not of immediate concern.

John bestirred himself, sticky and disheveled, looked at his watch, saw the hills, and soon he was lugging his suitcase through the station gate into the streets of Goslar, a beautiful gingerbread town that had fortunately been spared the ravages of war. Rather, its last direct experience of warfare had been very long ago, when wars were less destructive than they have been of late. Some buildings had construction dates from before 1600 carved on their lintels.

The lodgings that John found for himself were a mere fifteen minutes' walk from the mine, a short distance out of Goslar on the Clausthal road. The house was perfectly quiet and spotlessly clean. The landlady was an elderly skinflint. If she had a husband, he was so meek as to be to all intents and purposes imperceptible. One other lodger lived there, a permanent resident of great age, uncertain background, and ghostly habits, who confided to John in a moment of unaccustomed verbosity that he had worked in England as a waiter in 1913. This arrangement provided John with a place to lay his head in return for a modest weekly quota of Deutschmarks, and nothing more. Whatever he needed in the way of food, he had to find for himself.

John lived in a pale, airy room opening onto a balcony whose railings were brushed by fronds of the surrounding pine forest. He would leave to go to the mine in the cool air of dawn, cross the road, and scramble up a steep bank, pushing his way between saplings of fir and spruce. He would leave tracks in the dew-wet meadows, from whose margins deer melted into the surrounding woods, before following a zigzag path down through the trees to where he could look up at the mine buildings on the slope of the mountain. Sometimes, too, he would be on his way home soon after daybreak, or would walk from town up the road to the mine at dusk, for the production crews worked on three shifts around the clock.

The day shift began at 5:30 A.M. and ended at 1:30 P.M.; the afternoon shift ran from 1:30 to 9:30, and the night shift from 9:30 until 5:30 the next morning. By common consent the night

shift crews worked without a break until 3:00 A.M. At that time
they gathered in their lunchrooms underground, ate, and then
slept, wrapped in their coats, until 5:00, when they awoke and
stumbled, cold and sleepy, back to the shaft to catch the cage.

In the old German tradition, the miners at Rammelsberg went
about their work in a civilized, dignified manner. The mine was
an eye-opener to John, who, to his regret, never encountered
anything like it again.

A miner going to work would enter the mine buildings
through an arched gateway leading under the mine offices into
a cobbled courtyard. He then passed through an entrance hall
almost of the size and style of a nineteenth-century railroad
terminal, with a floor of polished tiles, decorated with murals
suitably indicative of the work ethic. The offices of the mine
officials opened off this hall. Passing through, he would enter
the dry, another bright, airy hall where mine clothes hung on
grappling hooks hoisted up to the ceiling on pulleys. Beside the
door leading out into the mine yard was a faucet that dispensed
tea of a sweet, fruity flavor with which the miner could fill his
thermos.

The company issued each man all his necessary protective
clothing, including a uniform suit of jacket and trousers made
of white canvas. When the suit was dirty with the black sulfide
ore of the mine, the miner handed it in, and it was laundered
for him and mended if necessary. The lamp battery fitted into a
box that hung low on the miner's back from a pair of shoulder
straps. This was more comfortable than slinging it from a belt,
as well as being less confining to those whose waists were dimensioned by the consumption of good German beer.

A canteen among the mine buildings served a cooked meal
in the middle of the day, plain, wholesome, and plentiful. A
man coming off the day shift, or about to go on afternoon shift,
could eat as much as he wanted for a few pfennigs.

The lamproom was built against the side of the hill. Looking
up as he crossed the mine yard, a man would see the mill building rising in steps up the hillside, tier upon tier, and above it
the headframe of the so-called Rammelsberg shaft, which was
used for rock hoisting. After collecting his lamp, the miner
would walk underground through a tunnel to the top of the
Richtschacht, which was the man-riding shaft of the mine. The
top of this shaft was underground inside the mountain; all necessary installations had been carved out of the rock. The shaft

was circular and lined with brick in the manner of a coal-mine shaft.

Most of the work was in progress on two levels, nos. 10 and 11. As the mine was beneath a hill, "depth" was a relative term, but 10 level was about 1,100 feet below the tunnel portal, and 11 level was about 130 feet below that. No. 12 was a tramming level, where ore was hauled from the stope chutes to a loading station at the bottom of the Rammelsberg shaft.

The active workings were almost entirely concentrated within the ore body, which in this vicinity was about 1,300 feet long and 250 feet across at its widest point. Back in the 1920s, the miners had found the ore body growing ever wider and longer as it went deeper. They were understandably optimistic that the mine would go on forever. It was therefore a bitter disappointment when they found that the thickness of the ore body resulted from geologic forces that had rolled it up on itself. Beneath that great bottom lobe they found nothing but barren slate.

A reliable axiom in looking for ore is that the best place to look for a mine is near other mines. Logically there was no reason why the ore at Rammelsberg should be unique. Unfortunately decades of searching failed to find another Rammelsberg, and the bottom lobe of the ore body merely ensured that the mine went out in a blaze of glory. Its extraction lasted for fifty years.

The mine was dry and pleasantly warm; the workings were spacious; machinery was used to accomplish much of what was accomplished by muscle power at South Crofty. Parts of the workings were permanently lit; there were even floodlights in the stopes.

On his first working shift John was sent with a quiet, sour little man to hoist equipment from an abandoned stope. John started looking for a rope, reckoning that the man would station himself in the stope, John on the level above, and they would have at it. Instead, they spent an hour or so finding a compressed-air winch mounted on a flat car and fixing it in place for use, and then they went to work. To John's everlasting surprise, very little obscene or blasphemous language was used because nothing seemed to go wrong enough to justify its use.

Centrally located on each level was a complex of stub tunnels, whitewashed and electrically lit, comprising a lunchroom, a shiftboss office, and a tool crib. Any tool or widget that a man might need was hung up or stored on shelves. A telephone system was connected to various places in the mine and the surface plant.

The lunchrooms were clean and well lit, with benches along the walls. The one on no. 10 level even had tables. Crews gathered there at the start of the shift for a bite to eat and a cup of tea or coffee. Shiftbosses would come in and call out the names of their men, and who was working with whom and where, and then everyone would troop out to start work. Halfway through the shift, they would gather once more to eat lunch. Supposedly, regulations forbade alcohol underground, but no one could see why a man should not enjoy a bottle of beer with his meal; if he abused the privilege, he was not worthy to be a miner. That was how things were done at Rammelsberg.

The ore body lent itself to operations on a more generous scale than did the narrow lodes at South Crofty. In the deepest part of the mine, stopes were laid out across the ore body from side to side, with unmined pillars of ore between them. These stopes were mined by a method known in English as cut-and-fill. As ore was blasted down from the stope roof, it was removed from the stope and then replaced with waste rock fill, which provided a new working floor.

For a month the floodlit hall of a stope would reverberate to the hammering roar of rockdrills as miners drilled hundreds of blasting holes ten feet into the hard sulfide ore of the stope roof. The holes were drilled upward, and the drilling water, laden with a black sludge of drill cuttings, spewed out all over the drillers. Drillers were unmistakable at the end of a shift.

When a stope had been drilled off, the shot holes would be loaded with explosives in batches of fifty at a time and then blasted, until the whole slice of ore, ten feet thick, lay in rubble on the floor, filling the stope to a depth of twelve to fifteen feet. A thin slice would be blasted off the roof with horizontal drilling, as this gave a smoother and safer rock surface. Finally miners would bar down any loose rock that might be hanging in place and would drill anchor bolts where necessary to secure any slabs that looked as if they might work loose.

The next job was to move the broken ore into steel-tubbing "millholes," which were built up as the stope progressed, so that they ran like chimneys through the waste fill. The ore went cascading down the millholes to chutes on the no. 12 level below. The machine used for doing this is known in North America as a "slusher hoist."

A slusher hoist is yet another of those devices specific to underground mining that is almost never seen in action on the surface. The hoist itself is a two- or three-drum electric or compressed-air winch mounted firmly in place. Tail-pulleys are hung from eyebolts drilled into the rock. The two or three steel cables on the hoist pull a hoe or bucket, vaguely resembling a bulldozer blade, back and forth over the surface of the broken rock that is to be moved. The bulldozing action of the blade moves the rock to where it falls down a millhole. The slusherman needs a certain dexterity on the controls to prevent the slusher bucket from falling down the millhole as well. By rearranging the tail-pulleys from time to time, the slusherman can scrape rock from wherever necessary. In a stope at Rammelsberg, the bucket might range over an area 100 feet long and 50 feet wide. The electric hoist would be of some twenty-five horsepower, driving a bucket four feet wide and weighing 400 or 500 pounds.

The work of a slusherman is seldom arduous, and at Rammelsberg this peaceful occupation lasted for a month at a time. The gradual disappearance of the muckpile down the millhole revealed the walls of the stope, beautiful in the soft floodlighting, with abstract murals painted by waves and folds and wrinkles in the bronze and gray ore. Indeed, this banded ore was prized for cutting and polishing to make household decorations. A miner was free to take as big a piece as he could carry. Often a man would look covetously at a boulder of particular beauty weighing a couple of hundred pounds and would shake his head sadly before sending it bumping and cracking down the millhole.

Once the pile of ore had been removed, it was replaced with crushed slate. This material was quarried on the surface, crushed, and dumped down a fill pass into the mine. Trammers drew fill from a chute and dumped it into whichever stopes were being backfilled at the time. The slusherman would spend another month distributing fill evenly through the stope. Once that had been done, drilling on the next slice of ore began, the miners standing on a level floor of slate.

In a stope at South Crofty a good crew would drill and blast almost every day, and the whole stope would be mined out in nine to twelve months. At Rammelsberg these "chamber" stopes, as they were known, followed a cycle of three to four months, and each one might run for ten years.

The massive pillars of ore between the stopes also had to be mined. A method had been devised that resembled an inverted cut-and-fill system. Miners blasted out the stope floor, which was solid ore, working under backfill that was elaborately supported on wire mesh and a mass of timber. The pillars had been taking load for years; the ore in them was cracked through and through, which made it extremely difficult to drill. Holes, once drilled, were often blocked by small fragments of ore and would have to be drilled again before they could be loaded. Water trickled through the pillars, which left the pillar stopes always wet and consequently made them undesirable places of residence.

The management knew that John was a mining student and saw to it that he received the best training that they could give him in the time available. Consequently, John never worked with any one miner for long. For a time he worked in the stopes, wherever needed, as a miner's helper. Then the company made him a "springer." Because the slushermen worked alone, the springer went from stope to stope, helping them as necessary and checking that no mishaps had befallen them. If no other work needed to be done, the springer ran the slusher while the slusherman smoked a cigarette. Wherever he went, the springer was almost always welcome. The job entailed some small degree of responsibility, as the springer was assigned work by a shiftboss for the first few hours of the shift, but from then on he was expected to plan his own work with minimal supervision and find his way around without getting lost.

By and by the company decided that John's education needed furthering and placed him and a Turk, formerly a policeman in the docks at Izmir, with Gerd Drohne in the training stope. John was to learn to drill and to load and fire blasts.

John's experience of "machines" at South Crofty had given him a jaundiced view of this kind of work. Traditionally, the machineman's mate did nothing but fetch and carry, hold the drill steel while the machineman collared each hole, and receive the copious abuse with which the machineman relieved his feelings about the hardness of the world. Using the machine for himself altered the picture entirely.

To be sure, learning the trade was not easy. At times the machine seemed to have an ill will of its own. From time to time the drill steel broke, and the machine hurled itself to the floor. The driller could either let go or be dragged with it; the former action was recommended as being less painful. Sometimes the spinning steel bored into a fissure in the rock and jammed, with the result that the machine would spin around on itself; the driller could only jump back and let it fight itself to a standstill. And, too, at times John could do no more than throw the machine down to rest his aching arms and wipe the sweat and black drill sludge off his face as he wandered over to talk to Gerd Drohne and Faik Sütsürüp, the Turkish policeman.

The Turks were a merry crew imported by a labor-hungry Germany. They established their own little communities in the towns where they worked. For the most part, they sent money to their families in Turkey and had no intention of staying permanently in Germany. They trickled in to Goslar to work for small businesses, where they were treated with varying degrees of callousness and dishonesty. One by one they congregated at Rammelsberg, as word got around that not only did Preussag pay better, but the Rammelsberg miners treated them like human beings. Notices in the mine were printed in German and Turkish. The Turks, hard-working and good-natured, had won their place among the German miners. With their disjointed lingo and irreverent sense of humor, the Turks often made the Germans laugh at each other and at themselves. After three months at Rammelsberg, John could claim with justification to be able to speak, besides English and German, pidgin German with a Turkish accent.

After a time under Gerd Drohne's instruction, John was turned loose as a driller in the chamber stopes. At Rammelsberg drillers drilled and slushermen slushed, moving from stope to stope as required. So John worked in most of the active chamber stopes at one time or another.

One stope was jinxed. Part of the trouble was that it came up under a crosscut on no. 11 level, with the result that the rock in the stope roof was cracked, but that was only part of the story. Most of the things that went wrong seemed to choose that stope to go wrong in. The jinx on that stope made a grab for John and nearly caught him.

John was on afternoon shift that week. One afternoon he ate

lunch in the mine canteen with some friends from the personnel office. With the bravado of inexperience—and not least, we may as well admit it, to impress a young lady whom he wished to impress—he declared that underground mining was not particularly dangerous. Within hours that remark would be conclusively disproved.

John went with Otto to drill in the jinxed stope. They were drilling shot holes inclined at seventy or eighty degrees, ten feet into the stope roof. The roof was about ten feet above the floor of crushed slate. Because of the angle of the drill steel, the driller was not standing directly beneath the rock into which he was drilling, but very nearly so. A more experienced man might have sounded the rock roof with a bar; sound rock will ring when struck, loose rock sounds drummy. The roof was, however, secured with rockbolts, and in places with wire mesh, and people had been working under it for the past two or three months, so perhaps a more experienced man would have done what John did—grab the machine and have at it.

To drill each hole, the driller stood the jackleg machine upright on its leg, squatted down to pick up a five-foot drill steel, and placed the shank in the machine's chuck. Turning the machine on so that it was just ticking over, he would cradle it in the crook of one arm while, with his other hand, he would adjust the leg valve to push the machine up toward the roof. The roof had been marked with paint for drilling, so he would try to stab a paint spot with the tip of the steel while the machine chugged and kicked and spat water into his face. Once the drill bit had cut a niche for itself in the rock by means of the driller's careful adjustments of leg valve and throttle, the driller would flip the throttle fully open, and the hammering roar of the machine would peal through the stope.

Squinting upward through the fog of the drill exhaust and the sludge spraying from the shot hole, he would try to keep the machine aligned so that the hole was drilled straight. Having drilled the five-foot steel to its full depth, he would turn off the throttle and leg valve, bring the machine down, and remove the steel from the chuck. It then took a balancing act to hold the machine upright with one hand and, with the other, pick a ten-foot steel off the floor, push it up into the hole, and fit the machine onto the projecting five feet of shank. He could then drill the rest of the hole, remove the machine from the steel, slide the steel from the hole, move the machine four feet back along the line of paint spots, and begin all over again. In that hard ore, fifteen holes was a good shift's work.

Toward the end of the shift, John had drilled a dozen holes, and his machine was hammering away at another one. In slow motion, a slab of rock peeled off the stope roof where he was drilling into it. In such situations a man develops powers of locomotion otherwise seen only in cartoon movies. At a distance of twenty or thirty feet he turned in time to see the slab, still in slow motion, land on the floor, bounce slightly, and come to rest. Appropriately enough, it was about the size and shape of a coffin. John had already drilled one hole while standing directly beneath that slab. Had it fallen then, it might have killed him. As it was, a smaller chunk of ore, about a one-foot cube, had landed precisely where he had been standing. Otto saw that something was wrong, stopped drilling, and came over to see what the trouble was. Finding John unhurt, he finished the hole he was drilling, and they decided that they had had enough for one night. They tidied the place up, busied themselves with odds and ends for a while, and called it quits.

Never again—ever—did John claim by thought, word, or deed that mining was a safe occupation. He recognized it for what it was—a hard, dangerous profession and, like the sea, remarkably unforgiving.

As summer drew on, John's thoughts turned, of necessity, to home. The miners urged him to stay with them. After all, they said, he would soon be ready for his Bergmannsprüfung (miner's exam), and then he would become a Hauer, a full-fledged miner with pay and privileges to match. John turned this idea over in his mind more than once as he looked up at the Rammelsberg in the early light of dawn, on soft summer evenings, or under the high sun of midday. But as before and as after, he chose the onward path, with its unknown risks and surmised rewards, in favor of the known attractions and known limitations of staying where he was.

He had to recognize, too, that never again would he be accepted in the same way as the South Crofty miners had accepted him as one of their number. To those men, their Cornish mines were the be-all and end-all of mining—indeed, the definition of what a mine was. As long as John knew no different, as long as his knowledge was their knowledge, his ignorance their ignorance, he could be and remain one of them for as long as he wished. Yet from now on he would always be in some measure a marked man, an outsider possessed of hidden knowledge, a wanderer of no fixed allegiance and calling no place home. With sincere and mutual regret John parted company with those who delved the Rammelsberg and mined its fabulous ore.

6

CANADIAN GOLD

The metal mines of Europe are old beyond living memory, in some cases old beyond recorded history. Therein lies their fascination. They exist, too, in areas of dense and ancient settlement. Miners are at work in Cornwall today whose families have been miners for ten generations; those who delved the Rammelsberg were heirs to forty generations of their predecessors. But therein lies the problem; these industries are long past their prime, and in many cases their continued survival is doubtful. In the 1970s it was the mines of North America and Australia, where new mineral deposits were still being found every year, that offered any adventurous young man the scope to exercise his talents to their fullest extent and to follow his calling without hindrance.

The mining country that called to John—for "call" is the only word to express the irrational attraction of a place about which he knew so little—was Canada. At the time of writing, Canada hosts one of the biggest and most technically advanced hardrock mining industries in the world. In the 1970s one in every six Canadians earned a living in mining or one of its associated industries and services. It is a little-known fact that at this time the mining industry of northern Ontario consumed more equipment and supplies than all the metal mines of the United States combined.

Someone who has not lived in Canada might imagine the country in terms of the Rockies, the Arctic, the Prairies, and cities such as Vancouver, Toronto, and Montreal. This mental picture probably would not include the most extensive geographic unit of all that is known, vaguely, as "the north" or "the bush."

This northern wilderness is so vast and so alien to common

experience that we must take time to examine it, for it is the stage on which many of the actions of our tale are played out and forms the somber backdrop across which flit the shadows of the actors.

If pressed at this point for a definition of "the north," we would have to say that it coincides with a belt of boreal forest stretching from the mountain country of the Alaska Panhandle, northern British Columbia, and the Yukon for 2,500 miles eastward to Labrador on the Atlantic coast. Its northern limit is the "tree line," a zone rather than a definite line, where even the fewest and most stunted trees give way to the tundra—the "barren lands." The forests grow taller and denser to the south, where their margins are bounded by the Prairies and by the farm country of the American Midwest, southern Ontario and Quebec, and New England. South of the Canadian border, the topography and other features of the north are represented only in the northern parts of Wisconsin, Minnesota, and Michigan.

The north is almost entirely flat, except at its eastern and western margins. Much of it is underlain by an immense slab of rock, ancient beyond comprehension, known as the Canadian Shield. The Canadian Shield gave rise to one of the greatest mining industries the world has ever seen.

Come with me. Let us board a train in Toronto in the dead of winter, bound for Winnipeg, Manitoba. The thirty-six hours needed to cross Ontario will give us the opportunity both to observe and to reflect on what manner of country this is.

We leave Toronto just before midnight. For a few hours, the train glides through rolling farmland quilted in snow. The lights of farms and small towns spangle the darkness. Before drifting off to sleep, we may remember passing through Parry Sound in the early hours of the morning, with which our journey is launched into the wilderness of northern Ontario.

We awake at dawn to find our train standing on a switch. We look out at the bush—brown undergrowth, white snow, black rock—so close, yet so remote in mutual indifference and in the brevity with which our gaze rests upon it. An eastbound freight train rumbles by—four diesel locomotives spurting plumes of thin, black smoke from their blaring exhausts, half a mile of snow-encrusted boxcars, tank cars, and flat decks, two more yelling locomotives, another half mile of freight cars, and the caboose, trailing a cloud of snow.

Our train crawls into Sudbury, the nickel-mining city of 100,000 people. It is a dull morning, with a gray overcast spitting sleet. A raw wind chills us to the bone as we stand on the platform in partly frozen slush to catch a breath of fresh air. East of town is a moonscape of slag dumps and black rock, white patches in hollows where snow has been allowed to settle by the restless wind. To the north and west we may glimpse smelter stacks and the headframes of some of the mines. In the 1970s half of all the miners in Ontario worked in mines like Creighton and Frood-Stobie, Levack West and Falconbridge, the nickel mines of the Sudbury Basin.

We are not sorry to resume our seats in the train or to avail ourselves of an ample breakfast. We should sit back and relax, because we still have another twenty-four hours before reaching Winnipeg on the bald prairies of Manitoba. All day long the train runs westward through the bush.

West of Sudbury high, rocky bluffs and rivers brawling between ice growing from their banks give way to more subdued terrain. Elements of scenery repeat themselves in endless permutations, never quite the same but never greatly different. A snow cornice, brushed by the wind into the shape of a breaking wave, rests on an outcrop of black rock. A tree trunk wears a half-coat of snow on its windward side. A stand of ill-grown birches furs the land with a dark stubble. A muskeg, thinly wooded at its margins with leaning and stunted black spruce, stretches to the horizon. A frozen lake, its low shores a tangle of broken sticks, fades out of sight into the mist and falling snow. Other than the railroad, the wilderness is unmarked by human hands, and we know that, away to the north—beyond the trees, beyond the horizon—it stretches empty for a thousand miles and more to the shores of the Arctic sea.

Every few hours we pass a settlement of low timber houses painted in whites and grays, dull reds and smoky blues, even its subdued colors an exclamation mark in the monochromatic forest. The streets are of packed snow, stained brown with oil and grit. The town stands silent under the gray sky, mantled thickly in snow. Behind it is the stillness, the loneliness, the profound silence of the deeply frozen bush.

Dusk falls, and although we would like to look out, the darkness is complete. We talk to our fellow passengers, read, eat a leisurely supper in the dining car, stare at the ceiling. If we wish, we can banish the empty darkness with the temporary camaraderie of the bar car, but there are times to savor enforced idleness.

The train slows down as lights appear outside; it stops in a small town. Steam drifts up past the windows from leaks in the heating system. A loudspeaker utters some words in a language that is probably English but, through boredom and overuse, has degenerated into a pudding of syllables. The peroration stops on an unfinished note, as if the loudspeaker was struck dumb with surprise at itself. Two railwaymen stand talking on the platform, muffled against the ferocious cold. The question of our whereabouts is answered by the conductor marching through the train. "White River!" he shouts, "Anyone for White River?" We slide once more into the embrace of sleep. Soon after midnight, we awake as the train pulls into Thunder Bay, its rail yards, docks, and grain silos lit up like day. The train stands for an hour to draw fuel and change crews, then pulls out into the night.

Another dawn finds us near Kenora, gliding onward among the same snow-mantled forests and frozen lakes as yesterday. A few hours later, however, fingers of open country penetrate the forest. The land is flatter and drier here, and we emerge onto the tabletop of the Prairies. It is midmorning when we reach Winnipeg, thirty-six hours out of Toronto. The train will run for another twenty-four hours yet to Calgary on the western prairies. Beyond that, the passage of the mountains will occupy at least another twenty hours before the train reaches its journey's end on Pacific tidewater in dreamy, fog-bound Vancouver.

Winter rules the north country for nine months of the year and is never far distant for the remaining three. Between November and March no liquid water is seen out of doors. The ground is like iron and several feet deep in dry, granular snow that drifts like sand and squeaks underfoot. On a sunny day in January, the temperature may creep as high as ten degrees Fahrenheit, but, when the light fades from the sky, the still air will be fifteen or twenty degrees below zero at nightfall, and thirty-five or forty below by dawn. When the wind blows, it seems that we can walk round all four sides of a city block with the wind in our faces all the way. It feels as if our faces are being worked over with a cheese grater, and our eyes water with pain. They call it a lazy wind; it goes through us because it is too lazy to go around. Fortunately the wind seldom blows when the temperature is colder than fifteen degrees below zero.

A northern blizzard has to be seen to be believed; every inhabitant of the north country will see two or three in a winter.

The storm may last for two days. Snow-laden winds howl in the deserted streets. Snow blankets everything, drifting in white dunes without respect for persons or property. A five-minute walk to the corner store becomes a series of tactical rushes from doorway to doorway, shelter to shelter. The only signs of life are snowplows, their flashing blue beacon lights reflected off the snow, working day and night to keep main thoroughfares open. At last the paroxysm abates; the sun comes out, and everyone digs. They dig with bulldozers and backhoes, snowplows and snowblowers, scoops and shovels. Snow is cleared from doorways and driveways, paths and buried cars.

In April a tentative softness comes into the air as the interminable winter breaks. In this season of snowmelt, earth and sky merge into a gray continuum. Frost-shattered bedrock merges upward with a waterlogged tangle of tree roots and black soil. Water stands between the trees over frozen subsoil. Roots join to become trees and undergrowth immersed in the misty exhalation of the cold, wet ground. Treetops reach up into the mist and the falling sleet and rain.

One day it is summer. Summer in the north country is an interlude between two winters to show us what the ground looks like—water flowing, plants growing and flowering in an exuberant celebration of life. The lakes and swamps breed hordes of mosquitoes, while May nights are loud with the chirping of a myriad of frogs. By day, temperatures may reach 100 degrees. The bush dries out. By night, lightning flickers all around the low, flat horizon, playing on an open tinderbox a thousand miles across. Thunderstorms rage over the empty bush, unheard by human ear; a swath in the trees may mark the path of a tornado. Forest fires sweep the land for months on end, leaving trails of blackened sticks a hundred miles long. A pervading smell of wood smoke and a dullness in the light of the sun warn townsfolk of a fire that may be fifty miles out in the bush. Only ceaseless vigilance protects the settlements from the fires that have engulfed them in the past. In peat land the fires go underground and burn all winter long.

Anyone who can goes to the lake, the camp, or the cottage for those twelve weekends of summer. But even summer can be cold and wet; snow has fallen in every month of the year. On a rainy summer day, the air is heavy with the threat of winter, always ready to resume its dominance.

As autumn comes, snow begins to fall and melt. The mosquitoes have gone; the air is often crisp and bright. In town the bars are loud with tales of the camp and the hunt. Snow falls a little more, melts a little less. Although crisp, bright days may linger into October, eventually a blast of snow and bitter weather announces to the inhabitants that winter has arrived and that they must resign themselves as best they may to another six months of snow and ice, cold and darkness.

In this forbidding wilderness men found ore. Gold was found in small amounts on the southern edge of the shield at Madoc, Ontario, in 1864, and a small but rich silver mine was worked briefly at Michipicoten on the north shore of Lake Superior. Copper and nickel were found during the construction of the Canadian Pacific Railway in the 1880s at a place known as Sudbury, where an enormous mining and smelting industry grew up that continues to this day.

In 1904 rich and plentiful silver ores were discovered at a place first known as Cobalt Station, T. & N. O. R. R., better known as just Cobalt. Cobalt showed the mining fraternity what riches in precious metals might lie hidden in the wilds of northern Ontario. Like Sudbury, the mines of Cobalt were discovered through the construction of a railroad. In this case it was the Temiskaming and Northern Ontario Rail Road pushing northward from its junction with the Canadian Pacific at North Bay with the idea of promoting agricultural settlement on the Clay Belt of northern Ontario. Like any such town, Cobalt became a springboard for exploration deeper into the wilderness.

In 1908 and 1909 prospectors fanning out far to the northwest of Cobalt found gold in a low bedrock ridge that barely broke through the muskeg. Over the next few years discoveries followed in profusion. The new camp became known as the Porcupine. To this day no one is sure why.

In the Porcupine, 1911 is still remembered as the year of the fire. July weather conditions and the convergence of several uncontrolled bush fires conspired to cause a horrible disaster that engulfed the new camp and the mines in a storm of flame. Timber buildings vanished without trace. Steel mine structures and machinery were reduced to tangled scrap. People who took refuge down mine shafts died when the fire passed overhead and robbed oxygen from the air. Many who went out onto or into Porcupine Lake drowned when a boxcar full of dynamite standing on a switch beside the lake blew up, leaving only a crater.

All that summer reconstruction went on at a furious pace

across the acres of charred sticks lying all askew, the black mush of rain-washed ash, and the fire-bleached bedrock. Autumn came, and snow gathered among the rock scarps. Soon a white mantle lay over the ravages of the fire and all human endeavors. Only a stubble of burnt trees broke its smooth whiteness. No foliage absorbed the violence of the blizzards that howled over the wilderness. Only man, foolish man, scrabbled night and day for that golden impurity in the rock.

Tall headframes grew up among the trees: their names were known throughout the mining world and in the offices of financiers. They were good mines. The rock was firm and strong. They encountered no hazards of water or gas, which plagued some other mining camps. Above all, the gold was free milling, plentiful, and rich. Hollinger, Dome, and McIntyre-Porcupine were the three big mines in the early years. Pamour Porcupine became the fourth in the 1930s. Others came and went over the years—Broulan Reef, Paymaster, Buffalo Ankerite, Preston East Dome, Vipond, Coniaurum, Delnite, Aunor, and the rest. The towns of Timmins, Schumacher, South Porcupine, and Porcupine grew up to serve the needs of the miners and gradually supplanted the bunkhouse camps of earlier years.

All through the twenties and the Dirty Thirties the camp boomed. The depression came and went, but the Porcupine knew it not. The gold mines roared and thudded, day and night, 365 days a year.

On Christmas Day, the bunkhouse kitchens excelled themselves—the day shift was let out early, and the afternoon shift was allowed to go to work late so that the men could enjoy their Christmas dinner. Men on the afternoon shift went to church in the morning, and day shift men went in the evening—those who were so inclined. The idea of observing the Sabbath or Christmas Day came late to northern Canada.

Men streamed in from the depression-ridden prairies and the coal mines of Nova Scotia. Some of them had not eaten for days and had not seen a square meal for months. Every day 200 or 300 men came to the mine gates, looking for work. Every day, at a certain time, the mine superintendent came out, looked them over, hired two or three, and dismissed the rest with a wave of his hand. The crowd moved on to the next mine in their endless, shuffling circuit. Those who were hired on worked seven days a week, sometimes in dynamite smoke and lung-eating rock dust. If a man did not like it or could not take it, there were 300 men at the gate more desperate than he.

Yet, in spite of the unemployed men at the gate, if a man was good at ice hockey, he could get a job at the mine, and a good one at that. If he let the mine superintendent fool around with his wife, so it was said, he could improve his position at the mine, or at any rate his stope or drift would make a good bonus.

In those days, the Cornish were strong in the community. A Cornish choir went on the air with the first broadcasts from the new local radio station. The shiftbosses and mine captains were Cornish and ruled the mines of the Porcupine as they had ruled the mines under Camborne and Redruth. A native Canadian could hardly understand their speech, and few outsiders entered this Cornish oligarchy. Besides the bunkhouses and small towns, clusters of makeshift housing grew up around the mines—Little Italy, the Black Shacks, Cyanide Alley.

The Porcupine back then was a booming, roaring, wild little community, newly carved out of the wilderness, with tree stumps in the back streets. Short, hot, mosquito-ridden summers were followed by the snow, ice, and mud of the interminable winters. Spreading in all directions was the subdued landscape of the Canadian Shield, green and brown in summer, in winter a monochrome in gray.

The British Empire realized the need for gold in the biggest war in history, even if the Americans did not. While War Order L-208 of 1942 practically destroyed the U.S. gold-mining industry, Canadian gold mines, although short of manpower and materials, mined gold and in some cases tungsten, which occurred with the ore in minor amounts, and made machine-gun parts in their workshops.

After the war a new tide of immigrants made their way to the mine gates from the horrors of devastated Europe. Men from every nation, many of them well qualified in their own occupations but speaking little English, hired on as miners, wanting nothing more than peace and quiet in which to live their lives and raise their families.

The price of gold was fixed in 1948 by international agreement, but the operating costs of the mines were not. During the 1950s and 1960s Canadian mining boomed as it never had before and probably never will again, for the country was rich in metals of all kinds—iron, lead, zinc, copper, nickel, silver, gold, uranium, and others. Vast construction projects, like the St. Lawrence Seaway and hydroelectric developments in Quebec and British Columbia, demanded tunneling, rock work, and construction of all kinds in unparalleled amounts. High wages paid by base-metal mines and big construction jobs drew skilled miners away from the gold mines. One by one, the mines of the Porcupine closed, falling victim to rising costs, declining ore grades, and the epitaph of mines the world over—"production caught up with development."

By the early 1970s, only five gold mines remained at work in the Porcupine. Because they were perennially short of men, they received no more than the normal degree of obstruction from the immigration authorities in hiring foreigners.

John entered into this situation more by chance than by intention. In years to come, mining gold in the Porcupine would give him an incomparable apprenticeship that would be the cornerstone of his career. In one way or another, the mining of gold would permeate his life for the next fifteen years.

7

THE GOLDEN PORCUPINE

John's acquaintance with the Porcupine predated his employment at South Crofty. It began with a letter. At that time his letters applying for work in mines in Canada descended on that country like snowflakes. Most of the recipients never answered. Yet at one company no less a person than the general superintendent took the trouble to write in reply, commenting that "we have employed a number of your countrymen in the past with marked success," but to his regret, the company was unable to make use of John's services at that time. Naturally, when John, as a student mining engineer, was looking for summer work in Canada three years later, he wrote a single letter, and he received a single, affirmative reply. The general superintendent's name was Robert Perry; the company was Dome Mines Ltd.

"Air Canada's DC-9 jetliner service to Timmins" sprang into the hot August sky, wheeled over the packed rail yards, factories, warehouses, and surging road traffic on the western outskirts of Toronto, and set course for the north. The sky was cloudless, and John was able to look down at the checkerboard landscape of southern Ontario. This pattern broke up as farmland gave way to forests scored by roads and powerline cuts, veined with silver rivers, and speckled with lakes that glinted in the sun. Nearly an hour passed. In England, which John had left the day before, an hour's flight in a jet airliner would travel the full length of a country of 55 million people. Over the European mainland, it could cross half a dozen nations, each with its own separate geography, language, culture, and history. Here, in Canada, it crossed 400 miles of empty wilderness, which was itself only a small fraction of the whole.

The aircraft descended and flew low over the flat, blank land-

scape, bumping in thermals rising from the hot bush. Roads, a railroad, mines, towns, and rectangular patches of buff-gray mine tailings came into view. John had no time to take it all in before the plane was once more over the bush. Trees by the countless thousand blurred beneath its wings. An airfield perimeter flashed by, followed by the threshold of an asphalt runway on which the plane landed with a sigh.

As he deplaned, John had a minute or two to take in the summer heat, the low, flat horizon, and the hot, blue sky overhead before walking into an air-conditioned terminal building. He observed his fellow passengers as they waited for their baggage. Among them were construction bosses in blazers and white hard hats, lawyers or accountants in pin-stripe suits with briefcases, and a scattering of people who could have followed almost any occupation. Some had friends or relatives waiting to greet them with a handshake, a hug, or a slap on the back. Others, like John, just waited listlessly.

As soon as John collected his suitcase, a man in a cap and uniform almost seized it from his hand. "Cab or limousine?" he asked. "Limousine," replied John, without knowing what the difference was. The man went outside to a full-sized North American car, threw the suitcase into a baggage-stuffed trunk, and opened the back door with a smile, beckoning John to the last vacant seat.

They lost no time in leaving the airport behind them. The driver inspected his passengers in the rear-view mirror and asked where they wanted to go.

The "limo" tore along the airport road. The forest opened out into meadows beside a river; a floatplane lay moored at a dock. The city of Timmins came into view. It crouched on a low ridge, its outline broken by the towers and turrets of the mines. It bore a faint resemblance to a fortress, and it had a certain grimness about it. The limo deposited John at the bus depot, a tar-paper shack beside a dusty gravel parking lot where he sat on his suitcase in the hot sun, waiting for the bus to South Porcupine.

The yellow bus bumped and rattled its way around Timmins before setting course for South Porcupine past the rockpiles and headframes of the mines. Headframes were sheeted in with painted iron for protection against winter snow and wind and thus took the form of strange, ungainly towers of various shapes and sizes. Timmins, Schumacher, and the mines disappeared behind them; the road sliced arrow-straight through the bush.

Settlements that seemed close together when seen from the air were in fact several miles apart, scattered through the forest; in such flat country it was impossible to see more than a hundred feet from the road. Here and there the trees opened out to give a view across a swamp or along a powerline cut, but for the most part the road ran on through the bush, whose fragrance blew into the bus with the hot wind of their passage. An old lady with her shopping who conversed with the driver in a foreign language was the only other passenger.

South Porcupine came into view, a cluster of white houses with two church spires side by side, asleep in the midmorning heat. It seemed to lie in a shallow bowl, around which the bush rose up on all sides. A low ridge, forested with conifers, provided a backdrop of sorts under the vast Ontario sky.

The center of South Porcupine was much like that of any other small town in rural North America—false-fronted business establishments and a dusty main street festooned with telephone wires. A block or two from the center of town, the main street ended at the shore of a blue lake. John lugged his suitcase around for a while, looking for a place to lay his head. He walked under a sign proclaiming the Algoma House Hotel into a tall, narrow room extending far back into the building. A group of three men sat at a table drinking frothy beer from small glasses. The room was pleasantly cool after the heat and glare of the street and smelled of beer and floor polish. One of the three men identified himself as the landlord.

When John had explained himself, the landlord, a silent, paunchy young man, divested him of ten dollars and led him up a flight of creaking stairs. They walked along a dim corridor, paneled with varnished wood and floored with dark green linoleum, worn and cracked, but polished so that it gleamed in the light from a lace-curtained window at the far end. The landlord showed John into a room, gave him the key, and left him to his own devices.

The room, like the rest of the interior of the building, was of varnished wood, well worn, but polished and spotlessly clean. A window with lace curtains looked out onto a greenish-black tar-paper roof. By craning his neck, John could just see treetops over one corner of the neighboring building. A wire insect screen prevented him from leaning out. The bedstead was of black iron tubing; as John sat down, it uttered the scratchy sound of a horsehair mattress and the twanging of metal springs.

He sat and stared at the floor, overwhelmed by the strangeness of his surroundings, trying to arrange the size and emptiness of this country in his mind. Although people spoke the same native tongue as he, this country called Canada was more alien than any of the countries of Europe.

He went downstairs and out into the summer heat, wandered down the street, and found himself some lunch in a restaurant. The waitress was from Redruth, but since she had left at the age of two, her recollections were sparse. John set out on foot for the mine. As it was a Saturday afternoon, no one was about, other than a bored security guard in the gatehouse, who greeted John with slow-spoken courtesy and advised him to come back on Monday morning.

The more mines John worked at, the more involved the hiring procedure seemed to be. It was not until 6:30 on Tuesday morning that he went in search of his shiftboss, whose name had been written for him on a scrap of paper. John's feet were encased in a pair of massively ribbed rubber mine boots, which safety regulations obliged him to buy from the company warehouse and which would have been only slightly more uncomfortable if they had been made of reinforced concrete. All necessary protective clothing could be obtained from the company warehouse, but not one item did the company provide free of charge. John later estimated that it would cost a man $100 to buy all the gear needed to work in a northern Ontario mine at that time. At current wage levels and after income tax deductions from his paycheck, a man would have to work for more than a week just to pay for his equipment.

Bob Drynan, to whom John reported, was a friendly, elderly man with frizzy gray hair who spoke with an accent straight from Ireland but who turned out to be from the Ottawa Valley. John would come to know Bob well in the course of time. The Ottawa Valley is farming country; the metaphors that Bob applied to the description of his fellow human beings and their actions were bovine in nature. He had hired on at Dome in 1937, in the days when a drift round was drilled with Leyner machines on bar-and-arm setups and was mucked with shovels off a steel plate—sixty tons of broken rock loaded by hand into four-ton cars in the space of an eight-hour shift. He could remember when jacklegs and mucking machines were new and strange. He had been in the army during the war but had returned to Dome afterward and had been there ever since. He was a quiet man, and if he held any deep philosophical opinions, he kept them to himself. John seldom saw him laugh and heard him do so still more rarely, but when he did, it was at some quiet little joke of his own. Bob had apparently been a shiftboss forever and ruled the cold, dank upper levels of the mine.

The miners standing about the dry that morning were huge, slow-moving, slow-spoken men, rendered larger by their thick woolly clothing. They eyed John pensively. John soon found out why people dressed so thickly and spent an uncomfortable shift following Bob on an introductory tour of his territory.

The mine was extensive and the upper levels were bitterly cold. The 9' x 9' drifts wandered and branched through the black rock. They saw few people. From time to time they came to the rambling caverns of a stope to find a miner, all alone, drilling with a jackleg while his partner was otherwise occupied in some other part of the stope. Visible only as a fuzz of light through the fog from the drill exhaust, the man would become aware of his visitors. When he turned the machine off, its slamming roar would fade, echoing, into the cold darkness. There were long, slow conversations with Bob Drynan in accents from England, Finland, Italy, and all parts of eastern Europe, as well as from Canada, which arrived gradually at some mutually agreeable conclusion.

This tangle of cold, dark drifts, inhabited, if at all, by large, slow-spoken miners, turned out three times the daily tonnage of South Crofty with all its busyness, or of the Rammelsberg. The answer lay in machinery—scooptrams, drilling jumbos, waggon drills, and Long Toms—all diesel and hydraulics and compressed air, which dug their steely fingers into the rock, tore it out, and chewed at it in huge mouthfuls.

"Summer students" were employed wherever they might be needed, without regard to any professional interest they might have in what they were doing, so John worked hard wherever he was assigned. He worked with a stope crew building timber fences to impound the sand fill that was piped down from the surface as a liquid slurry. He pulled chutes and trammed ore. He ran a jackleg with a wizened little Pole; they swore at each other furiously and became the best of friends. For one interminable week he worked with a longhole driller.

Soupbone was working by himself on the 700-foot level. The

level had long ago been abandoned except for the preparation of a blasthole stope. Soupbone was drilling blasting holes up to sixty feet long fanwise around a length of drift, in accordance with drawings given to him by the mine engineering department. When the time came, blasts involving several tons of explosives would send thousands of tons of ore cascading into a cavern to be drawn off 200 or 300 feet below. While the drilling was in progress, someone—no skills required—had to stay with the driller in case of accident and to help him move the massive rockdrill when necessary. Each ring of holes could be drilled from a single setup; each successive hole required nothing more than swiveling the machine on its mounting. When each ring was complete, however, the machine and its mounting of screw jacks, pipes, and clamps had to be manhandled four feet along the drift and set up again to drill the next ring.

One end of the drift was barricaded with a few planks; just beyond the barricade the drift ended in a black void. Soupbone threw a rock out into the darkness; a long time passed before it landed, with an echoing crack far below. The drift itself ran through the roof of this cavern. Great slabs of rock had peeled off, so that when Soupbone drilled into the floor, the drill bit sometimes broke out into the cavern beneath them after passing through as little as six feet of rock. Standing on a mere sill of rock, probably riven with cracks, which was being pounded by the machine, did little for John's peace of mind. Soupbone, with an unfailing but fatalistic sense of humor, waved his fears aside.

Center-stage in these proceedings was the longhole drill. The machine itself was three feet long and nearly a foot in diameter. It weighed, perhaps, 180 pounds. It was mounted on a slide with an air-driven screw feed. The slide was clamped to an arrangement of well-rusted steel pipes screw-jacked between floor and roof. When in action, the steel beast bellowed its demented fury at the rock, its exhaust port a single rectangular nostril continually exhaling, wreathed in a snot of oil and ice. A gray slurry of water and mud gushed steadily from the shot hole as the tungsten carbide bit on the end of the string of drill rods chewed its way into the rock. The rock was black and absorbed whatever light from the men's hat lamps penetrated the foggy drill exhaust. The sun, the wind, and the broad Ontario sky were half-forgotten figments of the imagination down there in the drill drift, filled with fog and thunder, darkness and bitter cold. The shattering roar of the machine shook the air and even the rock itself, blotting out thought and creeping into the innermost crevices of the mind.

John had nothing to do except pass four-foot extension rods to Soupbone; even that was unnecessary. When they moved the machine, he sweated; when the machine was in action once more, he froze. When Soupbone pulled the rods from a completed hole, the coupling sleeves were slightly warm from the pounding they had just received; John held them in his hands. He found a scrap of plank and stood on it so that his feet would not lose heat so quickly into the cold slime on the drift floor. His nose dripped, and he wiped it until it was sore.

Near the shaft was an electrically lit shack of timber and corrugated iron. The light bulbs made the place fractionally warmer than its surroundings. The two men thawed out there at lunchtime. Soupbone was a pleasant, talkative French Canadian, but he was paid for footage drilled, and after twenty minutes he would snap his lunchpail shut, and they would return to the cold and darkness and the thundering machine. He had no reason to teach anything to John; that would reduce his footage, and the loss would come out of his paycheck. So John stood and froze, marking time and turning slowly about. Minute followed minute, hour followed hour, day followed day of cold, intense noise and boredom. Nevertheless, John watched how things were done and stored the knowledge for future reference. After a week, Bob Drynan and his crew went on the night shift—the shifts rotated weekly. Soupbone worked for a contractor, not for Dome Mines, and stayed on the day shift. Some other unfortunate suffered the vigil of the drill drift.

John enjoyed the night shift. At night fewer people were in the mine because only production crews worked on that shift. People came to work in the hot summer evenings wide awake, fed and rested. Miners sat outside the dry and greeted each other with easy banter, then went slowly inside to change their clothes for the night's work. It was hot and sticky waiting for the cage in thick mining clothes, but they were soon underground in the cold workings of the mine.

The Dome headframe was so arranged that a man never went out of doors from the time he entered the dry at the beginning of his shift to the time when he left it to go home. A ramp led from the dry and the shiftbosses' wickets, up past a toolroom, into a room within the headframe about thirty feet above ground level. A doorway led straight into the cage. This arrangement was of small consequence in summer but was most welcome in winter.

At the end of the night shift the crew would walk out of the mine buildings at half past three in the morning into a balmy summer night. Sometimes the night was dry, and a hot, scented wind blew fretfully over the bush. Sometimes the ground was damp, the air cool and still, smelling of recent rain. Sometimes heat lightning flared soundlessly all around the horizon from thunderstorms far out in the bush.

Tisdale Bus Lines used to leave a bus at the mine gate. One of the security guards drove it on a round trip to South Porcupine and Timmins, returning to the mine an hour later. A spare driver would collect the bus in the morning. The security guards knew where their regular passengers lived and delivered them to their doors.

John would get off and stand alone in the quiet night. He would listen to the bus moving about South Porcupine and accelerating on the highway to Timmins. He would walk down to the lake to savor its stillness. With the first shiver of the predawn chill, he would turn back to his stuffy room in the Algoma House. Once in its hot confines, with the woodwork radiating heat absorbed during the day, he would take a bottle of tepid beer from a case that he kept under the bed and would read until his eyes were falling shut. Drawing the curtains against the first light of dawn and pulling the sheet over his head, he consigned the day to those who even then were forcing themselves awake to fight its battles, and fell asleep.

Soon summer was drawing to its close. Some of the winter residents of the Algoma House were coming in from the bush. In early September the weather was cool enough for deep sleep, but outside it was no time to hang about. Nights were frosty; by day the air was hard and startlingly cold under leaden skies. Old-timers said to each other, "Yeah, she's startin' to tighten up." Only those who have lived in the north can appreciate that expression, which means, truly, that winter is beginning to grip the land, although withholding for a time the fullness of its wrath.

At that time John was unaware of the strength of the Cornish connection in the Porcupine. The years from 1919 to 1921 were a time of bitter depression in the Cornish tin mines; the first waves of a stream of Cornish people immigrated to the Porcupine, which was then a new camp just overcoming the effects of World War I. Cornish mining skill, combined with American capital and engineering, launched the camp.

In the 1930s the Dome had been ruled by the Cornish; when Bob Drynan hired on in his youth, he had difficulty in understanding them. Now the real Cornish "old guard" were retiring, retired, or dead. Their children, and even their grandchildren, were working in the mines, often as craftsmen rather than as miners, but the old names like Moyle, Hocking, Uren, Stanlake, Libby, and Mitchell were all represented. One of the last of the originals was the assistant mine superintendent, Benny Mitchell.

During John's last week at the mine, just before one night shift, Benny Mitchell called him into his office. Spotting John's croust bag (most people carried metal or plastic lunchpails), Benny asked, "What have you got in that bag?" John replied, "That's my croust." The term "croust" is specific, not only to Cornwall, but to the mines around Camborne and Redruth. Benny Mitchell was from Redruth. So question and answer had a deeper significance than a mere inquiry as to the contents of a grubby canvas-webbing bag. Benny asked John if he would like to return for another summer job the following year. John replied that he wanted to try his luck around Sudbury; he was grateful for the offer, though, and would keep it in mind.

Just before leaving for England, John went to pay his respects to Robert Perry, whom he had not met. John discovered, not for the last time, that his work was observed by people of whose existence he was but dimly aware. Bob Perry offered him a permanent job after his graduation from the Camborne School of Mines.

8

TRUE NORTH

The first gold mines to be worked in Canada on any substantial scale were in the mountainous interior of British Columbia during the second half of the nineteenth century. Next, on a larger scale, came a string of discoveries along the Abitibi Greenstone Belt of northeastern Ontario and northwestern Quebec. Camps like the Porcupine, Kirkland Lake, Larder Lake, and Val D'Or were discovered between 1900 and 1930. Mineral exploration was not confined to gold, and several discoveries of base metals were made as well. When the price of gold rose in 1933, another wave of exploration found new mines to ward off some of the miseries of the depression. A new camp was found at Red Lake in northwestern Ontario, and another far away on the shores of the Great Slave Lake, only 300 miles from the Arctic Circle, that came to be known as Yellowknife.

Two mines were developed close to the town of Yellowknife—Con and Giant—which were of a relatively permanent nature. Others out in the bush, such as the Ptarmigan, Ruth, Tundra, and Thompson-Lundmark, were not. Since 1948, when the price of gold was pegged at thirty-five dollars an ounce, Canadian gold mining had not done well, as operating costs continued to rise. The Con mine was not one of the brighter jewels in the crown of the company that owned it and for years had been run on the proverbial shoestring. It was described by the managerial euphemism "well established."

The offer of a second summer job at Dome had its attractions, but John knew that there was more of Canadian mining to see before settling down to a permanent career, whether at Dome or elsewhere. John wanted to see what nickel mining at Sudbury was all about, but Inco and Falconbridge had their own supply

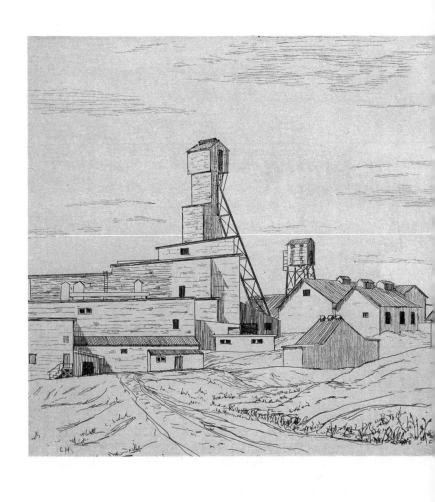

of summer students and had no place for him. By means that need not concern us here, John found himself headed for the Con mine at Yellowknife, that June of 1975.

The Pacific Western Airlines Boeing 737 sprang into the air, climbing fast over the Prairies, pale brown under the summer sun. From Edmonton, Alberta, they headed north for Yellowknife. Next to John sat a huge man in a bush jacket, drab green cotton work pants, and ankle boots, his leathery countenance topped by a gray crewcut; he was drinking out of a bottle wrapped in a brown paper bag. John bade him good day, but the man regarded John with a disdainful silence. John looked out of the window.

Patches of forest began to disorder the open prairie, which changed beneath them to bush country with clearings hacked out around small towns. These became fewer and finally ceased. The aircraft flew on over the wilderness of the Canadian Shield.

They came to a lake—more accurately, an inland sea. They left the shore behind them and floated in a luminous blue void, unmarked by cloud, contrail, boat, island, shoreline, or even a horizon line. They were in the middle of Canada out of sight of land, over a lake of whose existence most people were unaware, and whose name they had never heard. The Northwest Territories contains many of these—Great Slave, Great Bear, Contwoyto and Dubawnt, Kaminak, Kaminuriak, Kamilukuak.

Reaching the far shore, they made their descent. John glimpsed a town surrounded by a bare, rocky land. Trees and bushes grew only in hollows and gulleys in the seamed, gray rock. An arm of the lake slid by beneath them. A boat floated in the limpid water as though in space between the rock and the sky. A cluster of islands came into sight, along with gulls, a rock bluff with trees growing in a ravine, and a derelict headframe standing on a rockpile. Each color, each detail, was lit with startling clarity by the fierce subarctic sunlight. Floating down over a spread of gravel, the aircraft landed on an asphalt runway.

This was no international airport where garishly painted jets took tourists and high-priced executives to exotic destinations. It was an airstrip slapped down on the most convenient piece of flat ground for contact with the outside world and as a springboard for going deeper into the wilderness. It was a supply base for isolated settlements along the Arctic coast, drilling rigs, radar sites and weather stations, exploration camps and Eskimo villages. The afternoon flight from Edmonton was the descendent of the barge and the Cat train, and as eagerly awaited.

The taxi was an elderly vehicle that bore every sign of hard use on gravel roads. John could write his name in the yellow dust on top of the dashboard. The same yellow dust lay thickly inside the trunk, on the floor, and on those parts of the seats not cleaned off by clientele. Indeed, the only paved road within 500 miles was the main street of Yellowknife, and John was never sure whether it was paved or made of compacted oil and dirt. The same dust swirled in through the window with the summer heat in the course of their meteoric progress toward Yellowknife.

The stunted trees that swept past them opened out into the shabby outskirts of town. The premises of local trucking, drilling, and construction firms lay scattered about. The town was built on swells of bedrock, except for one development that had been built on muskeg. The houses had interfered with the freezing and thawing of the ground and now tilted visibly like toy houses on a rumpled carpet.

"Yessir," remarked the cab driver, "here's da #$%&* Yellerknife. One mad bad little town."

A black-clad figure reeled across the street. The cab driver put his elbow on the horn and swerved to avoid the vagrant. Leaning out of the window, he yelled at him, with an abundance of obscene gesture and hortative profanity, to "getcha $%&#$ ass outda @#$%&* road." Turning to John, he wiped his brow and remarked, "Gee, da drunken Indians is bad today," as if he might have commented that mosquitoes were plentiful on that scorching afternoon.

They roared out of town on a gravel road that corkscrewed over and around rock hillocks. Away to the right stood a white headframe among a cluster of buildings. The road swung to the left into a gulley leading down to the lake. A group of white buildings came into view. A stream of men climbed a flight of stairs into the largest of them. The taxi ground to a halt, scattering stones.

"Well, young fella, here's da bunkhouse. Yeah, five-bucks-fifty-take-easy-make-nice-eh." John's bags landed in a heap on the ground. The bellowing engine propelled the taximan's wheeled wreck back up the hill in the direction of town.

The Con bunkhouse camp consisted of timber buildings covered with whitewashed tar-paper; gray wood showed through flaking green trim. Apart from a few shacks of various shapes and sizes, there were three bunkhouses—"A," "B," and "C." They were two-story buildings standing directly on the hummocky bedrock and connected by wooden stairs and catwalks. "B" bunkhouse had only its bottom story in use; the top floor had burned out and had been boarded up and abandoned. Across a rock-strewn yard from the bunkhouses was the cookhouse. The eating hall occupied most of the top floor, thirty feet above ground level at the end nearest to the lake, and ten feet at the other, where the rock sloped up beneath it. A flight of steps led up the side of the building to the door.

John found a door in one of the buildings, crudely marked with the word "Offies," and looked inside. A man introduced himself as Louie, gave John an armful of bedding, and showed him into a room with two beds and a crudely made table. One bed was occupied by a supine figure that identified itself hesitantly as Billy Makinen. Billy was a young Finn who worked underground as a timberman; his brother worked at the Giant in a similar capacity.

John unfolded the bedding that Louie had given him. Patched and spotted sheets were stenciled with the company's name, as were the coarse, institutional gray blankets. John thought the bed uncommonly hard and, lifting the mattress, found a sheet of plywood underneath. On removing this, he found that the metal springs sagged a foot below the frame. As the plywood made the bed too hard for his liking, and the sag in the springs hurt his back without it, he bunched some of the metal links together and tied them with shoelaces. Billy Makinen watched with silent amusement. Straightening his back from fixing the bed, John wiped his brow and glanced out through the insect screen covering the window. Out there lay the subarctic bush, brilliant green under the bright sun.

"You going to eat?" he asked Billy Makinen.

"No, I eat already," replied Billy comatosely without removing his gaze from the ceiling.

John went across the yard to the cookhouse. A few men were still climbing the stairs; mostly they ignored him. The eating hall was a large, low room with tables and chairs around three and a half walls and a serving area along the remaining half wall. The walls were white but were smudged by the rubbing of countless shoulders. Worn and cracked linoleum covered the floor. In the center of the room was a table loaded with a rich variety of fruits, salads, cheeses, and desserts. Twenty or thirty men sat in various parts of the room, some in twos and threes, some alone. John recognized Louie, who, he discovered, was known as the bull cook, although his job had nothing to do with either bulls or cooking. Louie nodded to the people behind the serving counter to give John a meal.

John stared at the food in amazement. There was a choice of two main courses with soup, the salad table, fruit juices, tea, coffee, or any combination of those. If a man wanted three bowls of soup, he took three bowls of soup, like Louie's half-witted sidekick, who sat in a corner with an idiotic smile on his face, guzzling and burping over a row of soup bowls, for which reason he was known as Gravytrain. If a man wanted his plate stacked, the staff stacked it until it could hold no more. If he wanted another helping, it was there for the asking. For each meal some nominal sum came off a man's paycheck.

A rapid rate of turnover among the cooks ensured variety. One had borrowed money from the more gullible of the miners and then left town. One sold bootleg booze in camp on Sundays, when the bars were closed, until the security guards caught him. When one took potshots at the security guards with a BB gun, the cookhouse style changed yet again.

John loaded a tray for himself and wandered off, looking for a place to sit. Three young men sitting at a table beckoned to him as he walked by. Randy was a summer student from a university and was working underground. Larry was from Sudbury, also working in the mine. Mike was an Indian from northern British Columbia, who worked on the surface in the lamproom. John's account of himself brought raised eyebrows and approving nods.

John wondered aloud what work he would be doing. Randy replied, "They'll probably put you on tramming. Everyone does that. Stopes? They're mostly one-man, one-shift outfits. A lot of those guys don't like working with anyone else. Most of them have been here for years. Got it all sewn up. If you stick around for a few years, they might train you up and give you a try in a stope, but not just for the summer." John had hoped that his accumulated experience might count for more than hauling rock; this was not to be.

Randy warmed to his dissertation, being in the expansive mood that accompanies a well-filled stomach, picking his teeth between sentences to give each one its full weight. "You see, there's a bunch of company fat cats working in the stopes and drifts and raises, where all the bonus gets paid; then there's everyone else just coming and going. The trouble now is that the fat cats are getting fed up because their pay isn't keeping pace with the cost of living, and the price of gold's away up, so the company's making all kinds of money and not passing any of it on to them.

"The union contract's just about run out; I guess the company never told you there's a strike coming. Don't worry, it won't hit for a month or two yet. They're hiring and firing like crazy so as to get a bunch of guys living in the bunkhouse without two dimes to rub together, and hoping these guys will vote against a strike. What they don't realize is that most of these guys have nothing to lose and couldn't care less either way. What's more, they aren't making bonus like the fat cats, and they'll stand up at a union meeting and vote for a two-bucks-an-hour raise just like anyone else. If they get it, fine. If they don't get it, they pull a strike and head on out, leaving the fat cats holding the can and losing money.

"By the time they get a strike organized, it'll be fall. There's a month in the fall and a month in the spring when the Mackenzie River's freezing or breaking up where the highway goes across, and everything has to come in by air and costs like hell. That's what the storekeepers say anyway. Then the cold weather comes, and there's heating bills to pay.

"If there's a strike on, the fat cats will either tough it out and lose everything they've got, or they'll go back for ten cents an hour over what they're making now. Most of them have got families and payments to make on the TV, the camper, the boat, and they're going to get hurt. OK, so they quit and go to the Giant. Well, the Giant's the same, and their fat cats aren't going to let our fat cats in on the action, so they'll go tramming and busting rocks on the grizzley and making zilch. So nobody's going to win. We'll all be gone by then anyway."

"You want coffee, John?" asked Larry as he got up to get some for himself and the others. He brought back four mugs of coffee. The last mugful drunk, the last tooth picked, they handed in their plates. Mike asked the cooks for a plate of steak bones so that they could throw them off the landing outside the door and watch the lake gulls fighting over them.

The evening flight to Edmonton snarled overhead, climbing to the south. "Boy, I wish I was on that," sighed Randy.

"South of the border, down Alberta way," crooned Mike in parody of the song.

"Gee, I wish BC'd open up again. Place is shut down tighter'n the hinges of hell, and this place is drivin' me wacky," was Larry's comment.

John gathered that Yellowknife was not considered an ideal place of residence. The impression was confirmed by the stock conversational lead, "How long you got to do?" meaning, "How long do you have to work to raise the money to get out of here?"

They decided to go downtown to show John the sights. "Better put on some bug dope," they warned him, "mosquitoes like B-52s around here."

They walked the mile into town, where they made for the Gold Range. Better known as the Strange Range, it was a drinking establishment pure and simple, with no pretensions to any other function. It had no windows. The tables were covered with red toweling, held on by elastic hems, to absorb spilled beer. Many of the clientele wore hard hats. Weary waiters and waitresses, pallid in the harsh electric light, plied their endless rounds.

Three men were drinking at a table—to be accurate, two were drinking; a third was draped, unconscious, over a chair. Because sleeping in the bar was prohibited, the waitress refused to serve them. The two men picked up their comatose companion and slapped his face. He looked about him with a blank and startled expression and waved his arms. At this vital sign, the waitress went for more beer. The man stood up and walked out through the back exit, listing so severely to one side that it was difficult to understand how he remained upright.

Randy, Larry, Mike, and John had much to talk about. When they left at midnight, it was still daylight. They walked back to camp and went into their various bunkhouses.

John climbed a high rock beside the camp and sat down, looking across an arm of the lake to the low, rocky hills a mile away on the far shore. He needed to be alone to try to comprehend the strangeness of this alien land, harsh, raw, and crude, to which he must adapt, and whose challenge he must face.

It was nearly one o'clock in the morning. The air was still. John could hear mosquitoes humming all around him in the twilight. To his right was the sleeping camp. Turning to look over his right shoulder, he could see the white headframe of the mine, studded with lights, and could hear the thudding of the compressor. How strange it was to be sitting here on the shore of a Canadian lake in the midnight twilight, at a gold mine not far from the Arctic Circle! How far he was from where any of his forebears had ever been before!

John crept into the bunkhouse. Billy Makinen was asleep. He undressed quietly in the early light, climbed between the sheets, and fell asleep.

9

THE TRAMMERS

John left the bunkhouses in the bright morning to walk to the mine along a gravel road, dry and dusty in the summer heat. In places the rocky surface of the land was covered by shallow pools or patches of muskeg. Stunted trees grew in the shelter of crevices and rock scarps. Otherwise the land surface was of bare, gray rock, smoothed and rounded by the grinding of an ice sheet that had melted only a few thousand years before. From the crest of the slope, where the road came out of the gulley containing the bunkhouses, John could look back to the waters of the Great Slave Lake, which sparkled in the sun. This arm of the lake was only a mile or so wide, and beyond it lay the same subdued, rocky landscape as on the near shore. To the north, a mile away, were the outskirts of Yellowknife; ahead stood the mine.

The mine and mill buildings were ungainly tar-papered structures, white with green trim, clustered on a rock hummock around the headframe, itself white, but fouled with black grease that had dripped from the hoist cables over the years. It was said that the buildings had been constructed from wooden crates used to transport equipment, not only to Con, but to other gold mines in the camp, so thrifty had been the management, and so scarce the construction materials in those days. Indeed, the original stenciling—"NEGUS GOLD MINES LTD."—could still be seen on the unpainted interior walls of the mill.

Scarcely a single necessity of modern life could be found on this barren landscape. Every last nut, bolt, machine, tool, length of cable or pipe, pound of food or construction material, or gallon of fuel had to be brought in from Edmonton by one of three routes. It could be trucked for 900 miles along the Mackenzie Highway, of which 650 miles was dirt. It could be moved by rail

to Hay River and barged 120 miles across the Great Slave Lake, or it could be flown in. By the same token, it was not worthwhile to remove scrap machinery, which was abandoned at its last place of use.

The personnel office was among the mine buildings, and after going through the hiring-on procedure, John was left to his own devices for the rest of the day, with instructions to report for work the following morning.

Access to the mine was through the C1 shaft, running vertically from the surface to the 2,300-foot level. On that level a drift ran for half a mile to the top of an internal shaft known as the B3 winze, which ran vertically to the 4,900-foot level. The active workings at that time were all between the 3,500- and 4,900-foot levels, with the result that the journey to and from work included riding cages in two shafts with a mancar trip in between. Another winze, the C2, went on below the 4,900 to a depth of 5,600 feet from the surface, but it was abandoned and flooded. Unlike the bitterly cold upper levels at Dome, the active workings at Con were pleasantly warm.

John took his orders from a sourpuss Russian for some days before he understood that he had the privilege of working for this individual, a shiftboss known to the world as Slobodian. John nicknamed him the Humanoid and suggested to Randy that the man probably had a four-foot rockbolt for a spine, with its expansion shell anchored in whatever foul substance propped his ears apart. One day the corrosive juices of the Humanoid's interior would eat through the rockbolt, and Slobodian would disappear in a puff of pink smoke and nitric oxides. Randy laughed. "Yeah, you drew the short straw there alright. That guy hates everybody so much he even hates himself."

Slobodian habitually addressed John by a variety of vile epithets. John looked for the twinkling eye, the sly grin, which generally accompanied such modes of address; there was none. If Slobodian called John by some unprintable name, it was because he meant just that.

John saw Slobodian smile once. Before dayshift, the underground crew would lounge about inside the headframe, waiting for the cage, while a loudspeaker blared out the local radio news broadcast in Eskimo. One day, a dog trotted in and made the rounds of the miners, who petted it or ignored it according to temperament. John did not see what happened when the dog reached Slobodian, but he heard a sudden yelping and saw the dog departing the scene at a velocity approaching or equaling its critical Mach, throwing up dust and stones behind it. Slobodian had kicked it with his steel-toed mine boot in a delicate part of its anatomy and stood there grinning from ear to ear.

The other main feature of John's work was a skinny Norwegian, whom he nicknamed Eggbreth Eggbrethsen. Eggbrethsen had worked in the coal mines on Spitzbergen, among other places, and was employed as a trammer. John was assigned as his partner. Their job was to draw ore from stope chutes on the 3,700-, 3,900-, and 4,100-foot levels, haul it out to the shaft, and dump it down the ore pass. They would work one or more shifts on each level before moving to another level, and then to the third, or back to the first, according to which stopes were producing muck. One man drove the locomotive; the other "pulled chute."

On each level was a semiderelict electric locomotive, powered by a battery pack encased in a steel box. The assembled mass of corroded iron weighed roughly five tons. A headlight, a horn or bell, and brakes are considered desirable features of a locomotive. None of the locomotives on the three levels had horns or bells. One had a headlight that worked some of the time, but to compensate for that, it had no brakes. The method of bringing it to a stop was to engage reverse gear and jam a piece of drill steel under a wheel. This was known as a Hardmetals brake, after Boart Hardmetals, who supplied the drill steel. Each locomotive had a resistance coil in a perforated box beside the driver's feet. The coil glowed red-hot on most trips. Locomotives were known to catch fire from time to time. There were no fire extinguishers, and if the fire showed no sign of burning itself out, it would be doused with muck from the floor.

The locomotives pulled or pushed trains of six muck cars, each car holding about two tons of rock. The locomotive was always on the end of the train nearest to the shaft. Each car was dumped by overturning the body, which was built so as to roll over on its frame. The body was prevented from overturning while en route by a pedal-operated latch. Few of the latches functioned properly. Any mention of the mechanical condition of the equipment brought a blast of abuse from the Humanoid and words to the effect that if John regarded any aspect of the mine with less than complete approbation, he was welcome to pack his gear and hit the road.

The track leading to the stope chutes was idiosyncratic. It

had been laid by development crews who had driven the drifts many years before; they had been paid for driving drift, not for laying track like the Canadian Pacific Railway. Years of hard use without maintenance had not improved its condition. In places it wallowed in a bed of watery slime; each length of track rose up out of the water to greet the trammers as the weight of the train came onto it. Curiously enough, the way to avoid derailment was to drive flat out. The locomotive driver had to cling on to avoid being thrown off. He could only assume that the chute-puller was doing likewise, riding on the back of the bucking, lurching train.

At the chute the chute-puller worked on a platform about six feet above the track, using a bar to lift a steel arc gate and so release broken ore into the cars. The driver spotted each car under the chute by reference to chalk marks on the wall of the drift.

Not all chutes were built the same. The steeper the angle of the chute bed, the faster the flow of ore. The art lay in filling the car just right so that it would overturn easily at the dump. If the car was not full enough, it would be difficult to push the body over. If it was overfilled, rock would spill onto the track.

Eggbrethsen introduced John to chute-pulling on the fastest and meanest chute on the three levels. John filled the first trainload well enough, and the first two or three cars of the next, but then, as he held the chute gate open, a large rock slid down and lodged under the gate so that it could not be closed. The rock sat there obstinately while finer material poured out around it and filled the drift. They spent the rest of that shift, and part of the next, digging out the buried muck cars. John acquired an unwanted notoriety as "that new guy that can't pull chute." The Con miners had their own way of cutting strangers down to size.

Eggbrethsen continued this process by hazing John and calling him names. John fought fire with fire and muttered to no one in particular that Eggbrethsen's appearance might be improved by a pick sticking out of the back of his head, or simply yelled, "Shut up @#$% you!" with or without various addenda. Eggbrethsen quieted down, but John had to repeat the treatment from time to time. At lunchtime, they sat on the shaft station in a muck car body lying on its side and ate in morose silence.

Eggbrethsen was not alone in this ethos. One of the other trammers, also a permanent inmate, addressed his partner in

sepulchral tones with one of only two phrases, "To you partner," or "To me partner," indicating which way the locomotive
driver was to move the train. Luckless indeed were men sent
to work with a mucking-machine operator named Ladouceur.
Ladouceur's conversation consisted of nothing but vile abuse
from the start of the shift to its end. If anything went wrong, the
partner would find the verbal abuse augmented by rocks and
pieces of timber flying about his ears. Some men lasted as long
as five shifts.

No one liked to stand around the cold, drafty shaft station at
the bottom of the C1 shaft for long, especially at quitting time.
At the end of one shift, hoisting in the C1 shaft seemed to have
come to a complete halt. The miners, damp with sweat after
their day's work in the warm, deeper levels down the B3 winze,
were becoming restive. The shiftboss, who was there to maintain order, telephoned to find out what was going on. The miners fell silent as he replaced the telephone on its hook. "Guy on
the last trip," he announced, "didn't like the way the hoistman
ran the trip. Went around to the hoistroom and hit him in the
mouth. Hoistman said he wasn't hoisting no bunch of orang-
utangs like that and went home. Now they got to go find one of
the other hoistmen. Probably got to dig him out of the Strange
Range, so it'll be a while yet."

As John and Eggbrethsen hauled each train out to the shaft,
they dumped the rock onto a sieve of girders known as a grizzley. The rock piled up on the grizzley and had to be worked
down through it by hand. Oversized rocks had to be smashed
with a sledgehammer. The gaps between the girders were probably not quite large enough for a man to fall through, but the
question was academic, as there were no safety lines. The trammers of necessity acquired the habit of placing their feet in the
right places at the right times. Instead of being in a dead-end
drift, where the more obdurate oversize could safely be blasted,
the grizzleys at Con were directly opposite the shaft, and about
fifty feet from it. For that reason boulders had to be broken by
hammer, even if it meant wearing them down to dust in the
process.

When everything went well, they might pull as many as fifty-
four or even sixty cars in a shift. When things did not go well,
they might pull twelve or eighteen cars and sweat for every one
of them. One shift, they managed to pull sixty cars. When
Eggbrethsen reported this at quitting time, Slobodian abusively

accused him of false reporting. From then on, Slobodian's ration was fifty-four cars. In spite of his exhortations to "gif dat muck hell," that was all he ever got.

As mining went on, stopes were backfilled with sand tailings from the mill, which was piped down into the mine as a watery slurry. Each eight-foot-high "lift" of ore mined from the stope roof was replaced with an eight-foot layer of sand fill. When the next lift of ore was blasted down, the surface of the fill became impregnated with gold dust. For this reason, when the blasted ore had been scraped into the millhole with a slusher hoist, the stopemen would scrape out the top foot of fill beneath. When sand started to appear in the chute, the trammers knew that they were in for a tough shift.

John and Eggbrethsen began to pull sand at the end of one shift; next morning Eggbrethsen did not come to work. Slobodian sent John to pull chute with a new hire who had never worked underground before. The damp sand stuck in the chute; they barred it, hosed it with water, and finally had to blast it to get it moving. The jolting trip out to the shaft compacted the sand in the cars, so that it refused to come out at the dump. They had to overturn the car bodies, jam them in place, and then prod the sand with a bar while standing on the grizzley. Good judgment was needed to know when the whole mass was about to fall out, and so avoid being thrown down onto the grizzley under a ton and a half of sand. They pulled twelve cars that shift. Eggbrethsen knew that a shift's hard work would finish the job and was back at work the next day, saying, "See? I'm gone for a day, and my useless partner only gets twelve cars."

Slobodian knew full well what was going on. When his next pay slip came out, John found that he had been promoted from "Underground Laborer" to "Trammer," which raised his hourly wage from $4.04 to $4.09. He thanked Slobodian the next time he saw him. The Humanoid was not used to anyone thanking him for anything, so he glared and growled, shrugged his shoulders, and walked away.

One of the locomotive driver's jobs was to fetch stope miners from their work places at the end of the shift. A series of stopes followed one vein from the 3,700-foot level down to the 4,900. The ventilation air current flowed upward through these stopes. If crews in the lower stopes blasted at the end of the shift before they were supposed to, the crews in the upper stopes had to finish their work and climb out to the level in thick, choking smoke. More than once, John had to wait for the stope miners while smoke began to billow out of the manways from below. Quite soon, he could not even see the track from his seat on the locomotive, three feet above it. He heard the miners coming out of their stope and groping about. Guided by John's shouts, they found the locomotive, threw their gear onto the flat top of the battery box, and clambered onto the back bumper. "Keep your heads down, guys!" John yelled, and away they went, blind in the choking, gray smoke, until at last they came out into clear air. After such occasions, John felt ill all the next day.

For reasons that will appear shortly, one of the high spots of John's week was the Saturday shift. The miners worked voluntary overtime on Saturdays and Sundays to keep the mill fed. It is usually a lengthy process to start up and shut down a mine's ore treatment plant, so a mining company tends to run its mill continuously seven days a week. The mine works from Monday to Friday, delivering enough ore to the storage bins in five days to feed the mill for seven. Con did not work like that, and if at least some production did not come from the mine over the weekend, the mill ran out of feed and had to shut down.

John wanted to work on Saturdays for the extra money and because there was little else to do. At first he asked Slobodian for permission, which the Humanoid refused out of sheer cussedness. Then one of the men in the bunkhouse told John that he had only to pack his lunchpail and report to the Saturday shiftboss, as there was always work to be done and men missing. From then on, most Saturdays John worked with Scotty, which was always fun.

Scotty worked during the week with a timberman by the name of McCrimmon. McCrimmon looked as if his 105th birthday was imminent. His clothes hung in rags about his emaciated body. His toes stuck out through his disintegrating boots. His lunchpail was a mass of crumpled sheet metal held together with blasting wire. His voice emerged from somewhere beneath his hairy ears in a Scots burr so thick that only Scotty could understand him. McCrimmon needed Saturday to recover from the effects of Friday night. Sometimes the recovery process took until Monday. Eggbrethsen never worked overtime, so John and Scotty trammed ore together on Saturdays.

"It's more powerful than a jackleg, yet mightier than a mucking-machine! It's Super-trammers! Yeeeeehaaaaaaaaa!"

With these cries from Scotty, perched on the front of the locomotive, John slammed the throttle open. The wheels skidded on the track, jarring and grating, the locomotive snaking from side to side. Slowly the muck train gathered speed, leaving the lights of the shaft station behind them. Even in the main cross-cut, the track was as straight as a dog's hind leg. They lurched and crashed along, Scotty holding on for dear life. John had to keep one hand on the throttle, cling on with the other, and brace his feet to avoid being thrown off. Scotty yelled verses from "Snoopy and the Red Baron" with suitable sound effects and strafed ground targets, including the Saturday shiftboss on his rounds, who flattened himself against the wall as they hurtled by on their mass of mechanized scrap metal like two Riders late for the Apocalypse.

John soon knew every track joint on the three levels. On one trip inbound, the rim of the front muck car bounced just a shade higher than usual. John instinctively slammed the throttle into reverse, pitching Scotty headfirst into the ditch. Scotty picked himself up, laughing and cursing. Only the front wheels of the front car had derailed, which was better than having the whole train on the ground. Scotty soon had his own back. He rode the back car on the way out. They came to the shaft station; at least John thought it was "they," until he looked back and saw no one. He brought the train to a stop and walked around to the back, muttering to himself. There was Scotty with his lamp turned off, hiding behind the train.

Sometimes Hermann the German worked with them as a screen ape, busting rocks on the grizzley. They would come out with a full train to find Hermann still pounding ineffectually on a rock left over from the last trip. For all his size, Hermann was soft. Sweating profusely, he would vent his ill humor on the two trammers as if they were personally responsible for what came out of the chutes.

In true Super-trammer style, Scotty and John would seize a sledgehammer each and spring down onto the grizzley. With a few vicious and well-aimed blows, they would destroy the offending rocks and slot the pieces down through the grizzley. Scrambling back to the shaft station floor, Scotty would take his position beside the track; John would leap onto the locomotive. As each car passed Scotty, he would stamp on the foot latch and overturn the car body, which struck the steel crash beam with a resounding clang; its load of ore cascaded onto the grizzley. John would have the next car in front of Scotty before the body of the first one had righted itself. The last car dumped, he would reverse the locomotive and Scotty would spring aboard for the next trip inside. John would encourage Hermann in his work with some unprintable exhortation; screams of "Verdammter Scheisse!" and the sound of a sledgehammer being thrown across the shaft station would follow them as they headed for the chutes.

Scotty worked seven days a week for a month. Then one Friday evening John saw him in the Strange Range much the worse for wear. John joined him afterward for a nightcap in the room Scotty shared with Bill Bush. Bill was out, so John sat on Bill's bed. There was a fist-sized hole in the pasteboard wall.

"Who made that hole?" asked John.

"Mice, you fathead. Why don't you make another one?"

John punched another hole beside the first. Punching things was popular around the bunkhouse. Then they threw a flick knife at a pinup on the wall until John finished his beer and went to bed.

That was the last John saw of Scotty for nearly a week. About Wednesday, Slobodian asked McCrimmon where his partner was. McCrimmon stared at him as though at an imbecile. "He's been drunk for the past five days, you stupid bastard, so I expect he still is."

Repchuk, the personnel manager, visited Scotty's room. He found Scotty lying on the bed in a stupor. The cast-iron radiator was deep in a talus pile of broken bottles.

Repchuk surveyed this not unfamiliar scene. "What the hell is that?" he asked, pointing to a heap of splintered wood on the floor. Scotty focused with difficulty as he raised himself up on one elbow. "Well, partner, I guess it used to be a @#$%&* table. But I didn't like the @#$%&* color. Now why don't you @#$% off?" An empty whisky bottle smashed against the hastily slammed door.

Scotty returned to earth over the next day or so, eating delicately and looking white and fragile. He was soon his usual self, however, and the next Saturday, it was "Super-trammers! Yeeeeehaaaaaaaaa!" as they crashed along the disintegrating track, heading for the chutes.

BUNKHOUSE, BUGHOUSE

The personnel manager at Con, and to some extent anyone else interested in making the place run, lived under a cloud. The name of that cloud was "labor turnover." The statistical rate of turnover was 200 percent a year, which meant that, in theory, every one of the 250 employees quit or got fired and was replaced twice a year. In practice, however, this turnover was concentrated on the 80 or so men who lived in the bunkhouse camp. The rest of the crew lived in town or in a group of trailers between the camp and the mine and were relatively permanent in their adherence to the company (not that that adherence was based on any great degree of mutual regard). In the bunkhouse the coming and going was incessant; the place was regarded, not without reason, as a zoo.

Opinions varied. One man said that he had been living in camps for fifteen years, and "A" bunkhouse at Con was the wildest he had ever known. Yet from stories told of places like Thompson and Lynn Lake in northern Manitoba, or the stories Scotty told of camps in the Yukon and the more remote parts of the Northwest Territories, "A" bunkhouse was a sedate haven of peace and quiet.

For the most part John discovered little about who these transients were, other than his own immediate circle of friends. Either they were not there long enough to become acquainted, or something about them discouraged conversation. John saw some men every day at the mine or in camp, yet in greeting them, he would be met with a stony stare, if they even troubled to notice that he had spoken.

For some men, Yellowknife was a place to make money, either to leave it forever or to finance a grand debauch in Edmonton or

Calgary before returning to finance the next one. For others, the north was a refuge. Defeated, they pulled back into the wilderness, to the anonymity of the mine and the bunkhouse. They built impenetrable walls of privacy around themselves, their backgrounds erased. The stranger who tried to penetrate that reserve, however innocently, was not welcome—he was a spy, a hunter, to be feared and hated accordingly. As on any frontier, Yellowknife harbored a number of people who found it convenient to maintain a low profile in the affairs of this world. For many, hatred fueled their lives—hatred of their fellow human beings, the company, the mine, anything except the sheltering north. Perhaps the hurts done to them by the world had been deep, and they could not deal with it. Instead, they cursed it, turned their backs on it, and withdrew to a place where their physical needs were met and no questions were asked of them. Some of them were a little strange.

One of the mill workmen took a dislike to his shiftboss and, on some slight provocation, ran at him with an axe. The shiftboss fled; pursuer and pursued did several laps around the mill at a pace that would have won either participant a medal in Olympic hurdles. After a time, the shiftboss suspected that his pursuer was no longer at his heels. After putting in a few more laps for good measure, he ascertained that this was indeed true. The man was nowhere to be found. After two days he emerged from his hiding place up in the headframe—which directly adjoined the mill building—among the sheave wheels and skip dumps. They seized him and shipped him out to the lunatic asylum in Edmonton. After a short residence he was pronounced fit and returned to Yellowknife, where he continued his life as if nothing had happened.

"A" bunkhouse was the pit into which new arrivals were thrown; its internal life was bizarre. "B" bunkhouse, with the burned-out top floor, was quieter. When John had been at the mine for a month or so, Louie moved him into a single room in "C" bunkhouse. The personnel manager objected on the grounds that John was there only for the summer; Louie told him to mind his own business. "C" bunkhouse was occupied by those who liked their peace and quiet, so much so that John had difficulty waking up in the morning. Waking up was made even more difficult by a comfortable bed.

In "A" bunkhouse, where John lived at first, pounding feet and slamming doors heralded each day's arrival, while a variety of other sounds serenaded its departure. Of an evening, John and his friends might give the town a treat by going to wash their clothes, watch a movie, and drink a few beers. (There had been a washing machine at the mine, but someone had celebrated his escape by putting rocks into it and turning it on.) One of the bars was named the Gondola; the beer cost more, but the floor was carpeted, and the ambience was quieter than at the Strange Range. Often they frequented the Gondola. These things done, they would walk back to camp along the twisting, stony road in the midnight twilight with their laundry in garbage bags over their shoulders. Mosquitoes hummed about them, and the dust kicked up by their feet hung, luminous, on the still air.

The bars closed at one o'clock in the morning, and the heavy drinkers returned to camp some time thereafter. John never identified these creatures of the night. The washroom was next to the room that he shared with Billy Makinen. John was often jolted from sleep by a noise as if the wall was caving in and by the sound of breaking glass. Chancing to be awake one night at that witching hour, he heard someone stumble into the washroom, talking to himself. The man smashed his fist into the mirror and vomited abundantly into the sink. Louie would remove the shards of mirror the next day; the wall would remain vacant for a few days while he found a replacement mirror, and the cycle would repeat itself. The average life of a mirror was only two or three days; John kept a fragment to use as his shaving mirror.

Another man would come back to the bunkhouse, hooting drunk, and batter on a friend's door, demanding admission. After a time, he was subdued and pushed into his own room. The ensuing brief silence would be broken by a thud, followed by others in a crescendo, the last of each series being accompanied by the sound of breaking glass. John counted the events; the window in each room did not have that many panes. He concluded that the performer put empty bottles on the floor and stamped on them until they broke.

A tale was told that, back in the old days, when Yellowknife had been quite a wild little town ("before the Old Stope burned down and things got so #$%&* civilized around here"), a man wanted to visit his partner who lived at the far end of the bunkhouse. Rather than go there by means of the corridor, the man took the chainsaw that he kept under his bed and cut his way through the intervening walls. If anyone had any objections, they kept them to themselves.

In this volatile atmosphere disturbances were magnified out of proportion. One afternoon, John returned from the mine to find a full-scale shouting match in progress. It was Gerry— urbane Gerry, the philosophy student from an eastern university, who never swore, never raised his voice. Gerry was standing with Louie on the catwalk between the bunkhouses, looking at a cracked pane in the window of Gerry's room in "B" bunkhouse. Gerry had not cracked it and had no idea how long it had been like that. Louie was telling him that the nominal replacement charge would come out of his paycheck. Gerry waved his arms, shrieked abuse, and denounced the bull cook in terms that John had not even thought of. John stared for a moment, shrugged his shoulders, and went for his afternoon siesta. He knew the feeling.

John felt indeed that his life was controlled by lunatics. Slobodian was an obvious candidate. Slobodian and Eggbrethsen drank together, in spite of their professed aversion for one another. Louie had been widowed or divorced years before and had brought up his small daughter with lavish care and affection in his shack among the bunkhouses. Then the welfare people had come and taken her away, saying that he was not fit to look after a child. He had been slightly deranged ever since.

John and his friends were torn by irresoluble tensions. They needed each other's company, yet the same discussions endlessly repeated, the same gestures and turns of phrase, the predictability of each other, filled them at times with inadmissible hatred. They argued endlessly about whether there would be a strike or not. Each fresh scrap of information brought out the same chewed-over arguments to be chewed over yet again. They wanted to stay away from the mine on weekends, yet Sunday evening often found them tense with boredom. Every Sunday evening they would gather where someone had a television and watch "The Waltons." The Waltons were portrayed as quiet, sane, good people, and their influence on those around them corresponded to those attributes. That was why John and his friends watched the program before returning to the danger and lunacy of the mine.

Both mine and bunkhouse had a darker side, which the inhabitants could regard as tragedy if they so chose, or they could imbue it with a ghastly humor. For all too many, from the stories that John heard, Yellowknife was literally the end of the road.

A man went to work on night shift, late for work and drunk. The headframe was deserted. The drunk thought he could make up for lost time by sliding down the hoist cable. Either he took a dive for it and missed, or the water and grease on the cable gave him no grip. After falling or sliding for a great distance, he landed in the descending empty skip and was killed. When the skip reached the loading pocket, the skip tender, unawares, dumped a load of rock into it and rang it away. The crusherman on the surface was a little disturbed at the sight of a bloody corpse, covered in mud and dust, tumbling out of the skip dump, and ran all over the plant, hooting like a train. By the time the rest of the crew could get any sense out of him, the corpse had gone through the crusher.

Once, a man's lamp failed while he was underground alone. He groped his way out to the shaft station in total darkness. Unfortunately, not only were the shaft station lights out, but the gate guarding the shaft was off its hinges. The man stepped out and splattered the shaft walls with the revolting testimony of his demise.

One episode in the C1 shaft was never fully explained. The shaft bottom below the 2,300-foot level acted as a sump. The graveyard shift cage tender went down every night to check the water level. The hoistman would lower the cage slowly from the 2,300. A signal line hung down the bottom part of the shaft, and when the slowly descending cage touched the water, the cageman had only to pull sharply on the signal line to tell the hoistman to stop and bring it back to the 2,300 level. The cageman was not alone, but on this occasion his partner stayed on the 2,300 level. He left the shaft station on some errand. When he returned, he could see by the stationary hoist cable that the cage was stopped. The telephone on the shaft station was ringing. The hoistman was on the line; he said that he had lowered the cage to the track limit switch without receiving any signal and wanted to know what was going on.

The cageman's partner tested the signal line by ringing three bells—the signal not to move the cage—to which the hoistman replied by ringing three bells. (Signals sounded on a repeater buzzer on each shaft station.) The man then climbed down the shaft ladders. He found the cage fully submerged in the water, through which he could see the glow of the cageman's lamp. He scrambled back up the ladders and signaled to the hoistman to hoist the cage slowly to the level. The cageman's corpse lay on the floor.

The standard method of extracting gold from Canadian ore in the 1970s—as it had been for half a century—consisted of grinding the ore into a fine sand and leaching the gold out with a solution of sodium cyanide. Cyanide was commonly supplied as small, white bricks packed in fifty-gallon drums. A solution was produced by the simple expedient of knocking two holes in the drum and stuffing a water hose into one of them. The resulting outflow was the solution, which ran into a sump and was pumped to wherever it was needed.

Cyanide, in the form of either a clear solution or a white powder, is a deadly poison. At a mine in Ireland, cyanide showed up in a sugar bowl in the staff canteen and was prevented from carrying out its deadly mission only by the superhumanly sharp eyes of one of the mill engineers—but that is another story. At Con, a mill shiftboss thought that one of his men was taking a little long over his postprandial snooze. The snooze was permanent; there was cyanide in the man's coffee.

Con was fortunate that its relatively deep workings were comfortably warm. Giant, on the opposite side of town, suffered from the opposite condition and had to add antifreeze to their drilling water. Three thrifty miners thought that a purloined jug of this liquid would anesthetize their weekend in the bunkhouse. They were guilty of false economy; two men went blind, and the other died.

The old Norse idea of hell is a place of cold and darkness. Northern Canada provides these conditions in generous abundance. Winter, especially February and March, when the cold and darkness seemed interminable, brought some men in the bunkhouse to the end of their tether. An Indian blew his brains out with a hunting rifle. His room was splashed with blood and brains; the ceiling was blasted with bone splinters. Louie sent his half-witted partner, Gravytrain, to clean the place out. After washing the room down, he started to pick bone splinters from the ceiling. He wearied of this after a while and set to with a fresh coat of paint. The bone splinters remained to arouse the curiosity of anyone who noticed them. Because lying on the bed looking at the ceiling was a popular bunkhouse pastime, this ceiling must have given food for thought to a long succession of tenants.

But John had yet to experience the Canadian winter, and it was still high summer at Yellowknife. If there was one bright spot in the day, it was the view across the lake in the morning light. The lake shimmered in the sun, or veils of mist floated above its still surface. The true shape of the far shore was revealed little by little as the summer days went by. Here was an island, hitherto unnoticed against the gray rock hills, but disclosed when the sun shone from a certain angle. There was an unsuspected inlet etched by a patch of mist. The lake was forever beautiful, forever changing. From such a mere in Arthurian legend appeared the spectral hand that received the sheath of Excalibur.

Seen from the crest of the slope up from the camp, the headframe, although not an object of beauty by any stretch of the imagination, stood as a monument to the ingenuity and tenacity of this nation called Canada. But then the raucous shriek of the compressed-air mine whistle announced the start of the shift. It was a ghastly outburst, devoid of any redeeming qualities. It was a bellow, a scream, and a groan, combined in a vile discord that insulted the ear before fading into an expiring wail rolling mournfully across the rock and swamp to wash against the edge of town.

Billy Makinen and his brother departed by unorthodox means. They spent two months building a boat at the dock below the Con bunkhouses. The craft had a deep, V-shaped hull. John attended the launching ceremony and helped to ballast it with rocks. By the time the boat achieved any sort of stability, it drew six feet, and its remaining freeboard was alarmingly slight. With this craft the brothers planned to travel down the Mackenzie River—which was said to be only four feet deep in places—to Inuvik.

At Inuvik they would decide whether to turn left and sail around the coast of Alaska to British Columbia, or turn right and return to Finland. The unwavering certainty with which they proposed to take this ludicrous vessel through some of the wildest seas in the world left John, and probably others too, undecided whether the two men possessed hidden knowledge or were pathetically naive. They provisioned their boat and left their jobs, and John never saw or heard of them again.

Toward the end of August, nights became dark. The air was crisp at night; the mosquitoes disappeared. From time to time John saw patches of greenish light in the dark sky that waxed, moved about, and waned, indistinguishable at times from clouds lit by the moon.

Watching for the northern lights was like entering a theater without knowing whether the stage would remain dark and empty, whether a few actors would assemble in their street clothes to rehearse their lines, or whether a drama would be enacted greater than any written by human hand. From time to time patches of light focused into waving, green curtains, but one night, returning from town, John and his friends witnessed the aurora borealis in its full glory.

The lights were vastly still, yet never still. Incalculable forces threw gusts of light from the deepest horizon, far down the western sky, to the corresponding point in the east, yet in silence. Three curtains of green light hung across the sky; their lower fringes were white, turning to a purple hem. The curtains were made up of bars or pleats of light that lit up, one after another, so that a glow sprang across the sky, unhurrying, yet following each fold in each curtain at a hundred miles a second. They billowed like real curtains in a real wind, those curtains of light billowing in the winds of space.

After displaying their incomparable beauty for a time, the curtains faded to a dull glow. Some people say they have heard the lights rustling. The Inuit say they are the ghosts of stillborn children playing football with their umbilical cords. In any case, the northern lights are part and parcel of the strange fabric of human life in that bland, hostile habitat, the far north.

At Yellowknife John discovered one aspect of his profession that he had no wish to pursue. The company was having a new shaft sunk about half a mile from the Con headframe. A shaft-sinking contractor was doing the work. A representative from the company ensured that the job was done properly.

The shaft was being sunk on dimensions of 25' x 9' in rock and was now down about 3,800 feet on the way to a target depth of 5,800 feet. The arrangements were so designed that the shaft could ultimately be deepened to 7,200 feet. The hoist was to be a 2,000-horsepower monster imported from Germany, to be mounted in a tower that would be the tallest structure in the Northwest Territories. Rock would be hoisted in a fifteen-ton skip rocketing through the shaft at nearly 2,500 feet per minute. The whole job would cost about $12 million. It was quite an undertaking.

John and Gaetan, a mining student from Montreal, wangled

themselves a trip down the shaft one evening. The crew had just blasted a round in the shaft bottom and were sitting in the headframe, eating their lunch. Their blue hard hats and yellow oilskins were splashed with what looked like brown oil. The shaft leader snapped his lunchpail shut, at which signal they all stood up and shuffled toward the shaft.

The mouth of the shaft was covered by a pair of steel doors, on which rested a bucket six feet deep and five feet in diameter. The bucket hung by a yoke from the hoist cable; its rim was surrounded by a crosshead that engaged the shaft guides. John had seen photographs of shaft sinkers standing inside shaft buckets, but now, to his horror, everyone clambered up and stood on the rim of the bucket, holding onto the yoke.

The bucket trembled as the hoistman lifted it off the doors, which opened to reveal the yawning black shaft beneath them. John cursed himself for a fool and wished that he was securely seated in the Strange Range or the Gondola with a beer in front of him. The whole contraption sank into the darkness. The floor rose past them, and the doors closed overhead with a dull clang. The shaft crew thought nothing of it, laughing, joking, and lighting cigarettes. They all stood around the rim of the bucket, hanging on by their boot heels with six feet of empty bucket in front of them and thousands of feet of empty shaft behind them.

The shaft was framed and divided at twenty-foot intervals by steel beams fixed in place to support guides, pipes, cables, and other accoutrements. The bucket slowed to a standstill at the lowest and most recently installed of these. The company's engineer, some of the men, and John and Gaetan stepped onto the steel set. The two of them clung to the wire mesh that was bolted to the shaft walls to contain loose rocks. The miners made their way around the walls; one walked casually along a beam spanning the shaft, which floodlights now revealed as extending fifty feet below them to a pile of broken rock. Two men stayed on the bucket as it resumed its downward course, free of the crosshead. To John's horrified amazement, each man stood on the bucket rim facing outward, holding a scaling bar in both hands, tapping the shaft walls to detect loose rock. Only by pushing against the walls with the scaling bars did they prevent themselves from falling to the bottom of the shaft.

The bucket stopped when it reached the muckpile. After a few preparations, the clamshell bucket of the Cryderman mucker, mounted on a set of pneumatic booms, began to throw itself about, burping and hissing, loading rock into the bucket. The bucket was soon full and was brought back up to the steel set, where John and Gaetan and the shaft engineer climbed on. The engineer warned them to avoid kicking rocks off the full bucket onto the men below. They journeyed to the surface in the darkness of the shaft, silent but for the rustling of the crosshead on the guides.

Shaft sinking has always been the most hazardous of all mining work. Its hazards are intensified by the extreme urgency of completing the work in the shortest possible time. What the paratrooper or commando is to the infantry soldier, the shaft sinker is to the ordinary miner. The wages paid to shaft sinkers are often astronomical, but it need not surprise us that few shafts have been completed without someone being killed.

In this shaft two men had been killed in 3,500 feet of sinking. One man fell off the steelwork. The other man was newly arrived at this shaft. The operator of the Cryderman could not see directly beneath his machine. Therefore, by convention, everyone stayed on the clam driver's left when passing through this blind area. The new man walked around on the right. The clamshell smashed him against the wall. The total shaft bottom crew on all three shifts numbered fifteen men. At two fatalities among fifteen men in 3,500 feet of sinking, the odds of survival were not good enough for John's liking, and he avoided shaft sinking thereafter.

As the weather hardened and tightened with the onset of fall, so did the relations between the company and the union. There was more and more talk of a strike. A man going to work one evening was ambushed by men brandishing lengths of pipe. They told him it was as much as his life was worth to go any farther. He went back to the bunkhouse.

The union imposed an overtime ban, which put a stop to weekend work. No man in his right mind would defy the union, no matter what his personal feelings might be or what inducements were offered to him. John's attitude was colored by the fact that the union showed some slight degree of personal interest in his existence; the company showed none.

The mill crews worked a roster that allowed them to operate the mill continuously without overtime work, but in the absence of overtime work in the mine, the mill ran out of ore on Saturdays. As a result, the mill crew had to go through the tedious process of shutting the mill down and restarting it on Mondays when ore started to flow from the mine again.

One Friday evening a fault developed in the power cable to the B3 winze hoist. Because of the overtime ban, the electricians could not repair it until the following week, so there was no work (and no pay) for the underground crews on Monday either. John had only another week to work before returning to England. A bus was leaving for Calgary at midnight; he resolved to be on it.

He announced this fact as he joined the gang for supper in the cookhouse. Naturally, a farewell libation in the Strange Range was called for. Eggbrethsen was there and bought John a beer. "You know," said Eggbrethsen, "you were really good to work with." As they had exchanged little but commands and abuse, this surprised John no end. On the way out he noticed Slobodian sitting by himself, conveniently out of sight. Someone suggested that John throw a glass of beer in his face, but John had better uses for beer, and besides, he was leaving Yellowknife, while hapless wretches like Slobodian had to stay there, so he wandered over to say good-bye. True to form, Slobodian called him by a group of unprintable epithets, but to John's everlasting surprise, the Humanoid's sour face cracked into a grin. John and the gang went to a movie together and then had one last beer before they took him and his baggage to the bus depot.

It was dark and cold. A few flakes of snow fell on the little group clustered around the bus under the arc light. The passengers climbed on board, sleepy strangers. One or two well-wishers stood around, their breath standing with them in the still, cold air. The bus driver came out of his shack and mounted his throne. It was exactly midnight.

The engine revved, the brakes hissed explosively, and without any ado, the bus moved out into the night. John could dimly see his friends, waving, and they were gone. The last buildings and town lights passed behind them, and the bus went pounding on its way through the dark bush. It would be thirty hours to Calgary, and John soon fell asleep while the northern lights hung flaring in the Arctic sky.

THE TIN-STREAMERS

John's experiences in the mines of Germany and Canada were summer interludes between winters spent in Cornwall studying for a mining engineering degree at the Camborne School of Mines. The closer he came to the object of his studies, the less desirable seemed the consequences of its attainment. He had long foreseen that he would have to leave Cornwall, yet the longer he lived there, and as friendships deepened with the passing years, the more reluctantly he faced this necessity that every passing day brought closer.

John applied himself to the study of mining engineering as a temporary and disagreeable necessity with great diligence but scant enthusiasm. Paradoxically, the one thing that mining schools do not teach is mining. The subject matter for a "mining" degree imparts a broad theoretical understanding of mining and related engineering disciplines, whose value becomes apparent over a period of time measurable in decades rather than years, and then only if leavened by practical experience. If a mining school halved its expenditure on computers and similar devices and bought a dozen jacklegs, an Eimco, and a good supply of dynamite, it would turn out better mining engineers than is currently the case. The academic world, however, does not work like that; the only school of mining is the mine.

Being surrounded by mines and the remains of mining in every direction, Camborne School of Mines students absorbed a healthy element of practicality. (By contrast, the Royal School of Mines, situated in central London far from any mines, was said to turn out mere impractical pencil pushers.) One Friday, quite early in John's time as a mining student, someone yelled to him,

"Hey! Want a job tomorrow morning?" and John found himself with a part-time job as a laborer at a tin stream.

The ancient tinners, who washed cassiterite from river gravels, were known as tin-streamers; their works were called tin streams. Later, as underground mining developed, the streamers turned their skills to the waste products of the mines, which contained residues of tin.

The old Cornish mining industry plastered the landscape with tens of thousands of tons of waste products. Some of this material was barren granite and killas ("killas" was a term for the sedimentary rocks adjoining the granite). Some of it was low-grade tin ore that, at the time, was too poor to treat at a profit. Some of it was sand and slime discharged by the treatment plants, which had been piled into red dunes or which floored the shallow valleys with spreads of red mud. This ubiquitous red hue came from iron oxide, which permeated the tin veins. Mud from the mines, brick red when wet, would dry in the sun to rose pink.

These great quantities of mine waste provided numerous small entrepreneurs with premined, precrushed ore upon which to work their art and thus extract more tin. Most of the mines were abandoned in the late nineteenth century. The survivors owed their continued existence in some measure to their treatment plants, which were more efficient than their predecessors in gleaning as much tin as possible from each ton of ore, and which drove the streamers out of business as a result. Nevertheless, the waste piles from former operations remained, and when the price of tin was high, it was worthwhile to dig them up and reprocess them.

Brea Tin Ltd. was an assemblage of corroded hardware beside the stream in the valley that ran down to Tuckingmill. It was an enterprise whose periods of febrile activity were interspersed with periods of coma. There was a grizzley, a trommel, a ball mill, screens, pumps, and cyclones, all linked by pipes and conveyor belts. When in action, the whole affair revolved and rumbled, rattled and shook in a strange unison. A low building of corrugated iron sheltered a group of shaking tables, each about the size and shape of a billiard table, shaking back and forth with a ceaseless longitudinal motion. Ore that had been ground to a wet slurry was fed onto one edge of each table. This gentle shaking caused it to separate into its mineral and waste fractions, which splayed apart on the slightly sloping, riffled surface of the table and then crawled over the other edge and over the end of the table, to be collected in separate gutters.

The product of this grumbling collection of machinery, its interconnecting pipework squirting and dripping and bound up with wire and rags, was a black sand that was tin concentrate. A burlap bag of this concentrate, little larger than a loaf of bread, weighed fifty-six pounds. John found out all about those bags one Saturday morning when he and three other men loaded sixteen tons of concentrates onto a flat-deck truck to be taken to a smelter "up the line." John handled every one of those 640 bags in the space of three hours. At the end of the job all they could do was flop down, exhausted. The foreman slipped them some extra money that day to reward them for their efforts.

All week, while the plant was running, its conveyor belts dripped material onto the ground, so that piles of sand, slime, or gravel accumulated underneath them. This had to be shoveled out and thrown back onto the conveyor, or into the bucket of a front-end loader. As willing banjo artists were in short supply, Brea Tin Ltd. was happy to pay mining students in coin of the realm to clean conveyor spill on Saturday mornings. The pay was about three pounds sterling. At five pints of beer to the pound sterling, the money was all spent in the pub by Monday, so the business helped to fuel the local economy.

John would get up soon after dawn, sling his croust bag over his shoulder, and set off through the quiet streets of Camborne. In the winter months it was often raining or drizzling, but there were bright mornings lit by the early sun, the light of the new day gleaming on roofs and streets washed by the night's rain. Occasionally there were bitter frosts, and they would be lucky if the plant did not freeze up. From time to time it would be blowing a storm, and John would have to fight his way against the wind that ambushed him at street corners, hummed in the telephone wires, and set glass milk bottles clinking as it bowled them along the streets. Leaving the town behind him, he would follow a lane running up the valley from Tuckingmill. To his left were the buildings of South Crofty, the ruins of Cook's Kitchen, and the fingers of engine stacks pointing at the sky. To his right were the cottages of Brea Village. One was obviously the home of a retired tramp miner, with "Dunromin" painted on its neat little front gate.

On reaching Brea Tin, he would report to Don Pascoe, the

foreman, poke his head into various parts of the plant to see who else was around, and make for the lunchroom. The lunchroom was a windowless lean-to built onto one side of the plant building, lit through a panel of translucent corrugated plastic in the roof. Its only redeeming feature was that it was warm and provided shelter from wind and rain. It stank of dirty socks. The walls were decorated with pinups so ancient that dust had collected on the curls and wrinkles in the paper. The galvanized iron garbage pail looked as if it had lost an argument with a D9. It seemed to get emptied every six months, and the rotting refuse at the bottom contributed its own faint, sour odor. John and whoever else was there would have a quick cup of tea before looking for a long-handled shovel and going to work.

Mining school was a grind—mostly engineering theory, voluminous, and dry as chalk dust. Fresh air and exercise were necessary to clear the brain of its weekly burden of numbers and formulae. The banjo artists would work for two hours, shoveling conveyor spill, take half an hour for tea break, shovel for another two hours, and go home with a few pound notes in their pockets. One Saturday John forgot to take the money out of his coveralls before putting them through the laundromat. Maybe that was not what was meant by "money-laundering," but the few scraps of paper that fluttered out were not acceptable currency in the Plough—not that that prevented him from drinking his fill, perhaps with even more gusto than usual. John tried to estimate how much sand and gravel he shoveled in a morning from the number of front-end loader buckets he filled. The answer, one morning, came to fifteen tons. Whenever the plant reopened after temporary closure, John was sure to get a call for his services as the Saturday morning banjo operator.

Brea Tin was responsible for keeping the Tuckingmill stream, known as the Red River because of its burden of mine wastes, cleared of the trash that people threw into it. One May Saturday, John and Rullyon Rattray were given this additional duty after their morning's work.

They went down to the appointed spot and began pulling out old tires, corrugated iron, baby carriages, bicycle wheels, and suchlike. At first they tried to spot each piece of debris from the bank and then drag it out, but they were soon soaked to the waist, their boots full of water. It was simpler to wade downstream, feeling for debris with their feet. The water was warm under the first sun of summer and opaque with mud. In places the stream ran between high, gorsy banks, where they could climb neither in nor out.

Carefully, they would grope around the wreckage, trying not to tear their hands on jagged metal. Ten minutes of twisting and pulling might extract a piece of barely recognizable scrap from the sandy bed of the stream, to be thrown up over the bank. John and Rullyon knew that the evening's entertainment for the local youth would be to throw it all back again, but they were not paid to worry about such things.

The mines between Camborne and Redruth lay on and in a gently rolling plateau, overlooked by the northwest slopes of two granite hills, Carn Brea and Carn Entral. This plateau ran for three or four miles to cliffs overlooking the sea, cleft by the valleys of streams running down from the moors.

Back in the 1600s and 1700s, when the mines at Camborne and Redruth were first opened, mine drainage was so difficult that it was worthwhile to drive thousands of feet of drainage tunnels from the valleys in order to drain the mines to depths of 100–200 feet from the surface. These adits also served to explore the lodes. In some cases good success attended this work; in the eighteenth century, Pool Adit was one of the richest copper mines of the district. By modern times, most of these adits were lost, their portals collapsed and obliterated.

The most extensive adit system was the Dolcoath Deep Adit, driven in the mid-1700s, whose branches rambled for miles under Camborne and the villages of Tuckingmill, Pool, and Illogan Highway. Parts of it were maintained by South Crofty, whose pumps discharged into it. After a course of at least a mile downstream from South Crofty, it discharged into the Red River at Roscroggan Bridge. John wandered down there once, but an ill-natured watchman sent him packing.

John and Rullyon worked their way downstream toward Roscroggan Bridge, which was the limit of their task. John remembered the adit, so they waded on under the bridge to where the adit water joined the river after flowing from the portal for some 200 feet through a deep trench. If the watchman found them in the trench, they would say, innocently, that they had been told to clean the riverbed, of which their appearance would give ample proof.

Unlike the rushing, turbulent stream, the adit water in the trench ran clear and silent. A green canopy of tree branches arched overhead. The walls of the trench rose ten feet above their heads, but the watchman's cottage stood at its very edge, and the water made no noise to cover their movements. John touched the bank to steady himself; a small piece of earth fell into the water with a plop. They cowered under the bank when they heard movements in the cottage. They waded ghostlike through the dim, green light until they reached the tunnel's mouth, which had been barricaded with scrap drill steel. Whoever had done the job must have been very fat, because the two young men had no difficulty in passing through. The portal measured about six feet in height and width. After a brief look inside, they ghosted back to the main stream.

There being no time like the present, they decided to go into the adit that night. They had no time to borrow mine lamps, so three flashlights would have to do. It was not entirely a safe enterprise. The stream of water flowing in the adit drew air with it, so ventilation was not a problem. However, if South Crofty for some reason decided to run their pumps at full bore that night, anyone in the adit could soon be up to his neck in water. There was a faint chance that the adit might flood under one of the thunderstorms that from time to time split the summer nights. Finally, the three flashlights could fail, leaving the two explorers to wander blindly in workings two and three centuries old. But what of it?

Late that evening, after dusk had fallen, they drove in Rullyon's car to a place a short distance upstream from Roscroggan Bridge. They left the car under a hawthorn bush spangled with white flowers like small, dull stars in the fragrant night. They crept down to the stream, looked furtively around, and slid in like otters. They waded downstream, waist-deep in the cool water, and passed under the bridge.

The silence in the drain trench had been complete by day; now it was intense. They could have heard a mouse snore. The velvet darkness was impenetrable, and they dared not use their flashlights until safely—if that was the word—inside the adit. Their stealthy movement was yet quieter than before. John's heart was in his mouth when a bird clattered away through the leaves overhead. A piece of timber wobbled underfoot, threatening to tip him into the water with a splash fit to waken the dead. The sweet smell of rank vegetation hung on the still air. After a seeming eternity they slid into the tunnel. It was nearly midnight.

The cool breath of the adit greeted them with the musty smell of the few feet of earth overhead. They turned on two flashlights. The earthy roof and walls, propped with alarmingly few cracked and bulging boards, gave way to bedrock. The tunnel jinked left and right and headed for the mines.

At first they had room to spare between their shoulders and the rock walls, but they had to stoop. Their hard hats scraped the roof; they longed to straighten their backs. They came to an air shaft where they could stand upright and stretch their aching necks. Several times they stopped dead, whispering, "What was that?" But it was only their breathing, or their clothes rustling, or water flowing, magnified by their trepid minds. The water was between knee- and thigh-deep. The walls were coated with red mud; a high-water mark showed ominously close to the roof.

The adit became harder to pass through. In places it had an oval shape, perhaps as much as seven feet high, but inclined to one side, so that they had to slide along on their left shoulders. Even so, their shoulders were a sliding fit in the rock. The oval was so perfectly shaped that they could not put one foot down beside the other. They came to a rathole stretch, a mere four feet high and barely two feet wide. The water was two feet deep. They had to force their shoulders through, squatting in the water. The miners who drove that part of the adit must have been small men working on their knees.

From time to time, they came to air shafts with ladders leading to locked gratings on the surface. These were used by the men who inspected and maintained the adit. From time to time the adit crew appeared in town, incongruously dressed in oilskins, hard hats, and muddy thigh-boots, their lamps lit in broad daylight. Standing at the bottom of these shafts, John and Rullyon could hear traffic on nearby roads, and once they heard an aircraft droning through the night sky.

One flashlight went on strike. They agreed to keep the other in reserve. John was leading, so Rullyon had to push on in darkness relieved only by the chinks of light showing between John's body and the rock. The low roof crammed their hard hats down over their eyes; their necks ached from trying to look ahead.

They passed tunnels branching left and right—the shallow workings of old mines that had followed the copper-ore veins long ago. After an eternity of sloshing through the water, they came to a place where the adit branched into two and an air shaft ran away up to the surface.

A board was wedged between the walls, with an empty cigarette packet lying on it. They sat down.

"What time is it?"

"One in the morning."

"Where are we?"

"Don't know."

"Why are we here?"

"Don't know."

Their guffaws echoed down the tunnel, suddenly cut short by the thought that someone on the surface might hear laughter coming up the old shaft. Some of the shafts came up under houses, or close to them, and John had heard stories of voices and hammering noises beneath what seemed to be solid ground.

The ground under Camborne was a warren of old mine workings, mostly uncharted and forgotten. It was not unknown for a gaping shaft to appear in what had been a green field as far back as anyone could remember.

One man kept pigs in a sty behind his house—until he looked out one morning to find that the sty, and the unfortunate pigs, had been engulfed by a shaft that had opened up in the night.

Another man was persuaded by his wife to lay new linoleum in the kitchen. After ripping out the old material, he noticed a draft coming up between the floorboards. He lowered a nail on a piece of string to investigate this phenomenon. After letting out twenty feet or so, he realized that the kitchen was above an old shaft.

Having left John and Rullyon to their brief repose, let us now go with them farther into the Dolcoath Adit. The tunnel twisted like a snake, propped in places with steel girders and brickwork. It became wider and higher, and the stream shallower, so that they made better headway. They came to a long, straight stretch, driven in the 1950s to avoid a section that was likely to cave in. They could walk upright, and the water was only inches deep on a gravel bed. They tore along at a great pace, but the adit ran on, straight as a die, into the darkness.

They stopped. It was two o'clock in the morning. One flashlight had given up; the other had been on for two hours. They knew that if its battery died, then, by the workings of Murphy's Law, the third flashlight would also fail. Neither of them had the least idea where they were, how far the adit went, or how or where it ended.

From the seat board where they had rested, the adit ran directly to its portal. They had noticed that, where tunnels branched off to either side of that section, the water ran in a trench. Even in complete darkness, they had only to follow that trench to find the portal. Where they were now, however, as well as branches, there were flooded shafts and other things that were not good to encounter without light. They decided to retreat, which they did with all possible speed, and were relieved to return to the seat board and the narrow doghole that was the passageway to the surface.

They pushed their way back to the portal, slightly aided by the fact that the stream of water now flowed with their stumbling feet instead of against them. They emerged under the night sky at three o'clock in the morning. They ghosted back down the drain trench and waded under Roscroggan Bridge upstream to where the car was parked.

John brought Rullyon in through the back door of his lodgings in his sock feet, covered a chair with newspaper, and motioned him to sit down while he brewed tea. The hot, sweet tea was like nectar to their throats. The electric light in the kitchen revealed two apparitions, bone-weary, soaked with water and sweat, and smeared from head to foot with the red mud of the adit. Rullyon went home to his lodgings. John threw his clothes in a heap on the ground outside the back door and, after some rudimentary washing, crept upstairs to bed as the night paled with the first light of dawn.

John was fortunate that his landlord and landlady—Willie and Lizzie Symons—were tolerant of eccentricity. They treated him like a son for the three years that he lodged with them. In so doing, they helped him to earn his mining degree far more than they knew.

John's experience of Yellowknife had so disillusioned him with that part of Canada that he had written from there to Bob Perry at Dome Mines Ltd. to ask if the job offered the previous autumn was still open to him. The reply was affirmative, so that John began his final year at mining school with a firm offer of employment. The immigration procedure ground through its nine-month routine by fits and starts; by the time he passed his final examinations, John was in all respects ready to leave for northern Ontario.

The summer of 1976 was long remembered in England as one of scorching heat and drought. The English summer sun breathed its warmth into John's bones, and everything about Cornwall seemed to implore him to stay.

The School of Mines was just across the road from South Crofty. The northern crosscuts from Robinson's shaft lay directly beneath where John toiled with pen and slide rule. If he went back there, he would be among friends; Edgar had met him in the street one day and asked when he would be coming back. It would be so easy to cross the road, walk up the familiar lane, past the timber yard and the thudding compressor, and tell Ted Rowe, the personnel manager, that he wanted a job. Ted would squint up at him from behind his desk with a grin and reply, "So you come back, huh?" and that would be that—so easy.

If he did that, he might as well take his mining degree, on which he had spent years of toil, crumple it into a ball, and toss it over the hedge. John knew people who had stayed in Cornwall. They had progressed to a certain point in their careers but then had been passed over in favor of others with overseas experience. A man with experience gained in North America, Africa, or Australia might, in later life, have the option to return to Cornwall. The reverse option did not exist. In North America, in particular, it was North American experience that counted.

The Cornish mines lived and died by the price of tin, which had always been wildly unstable and was likely to remain so. In the early 1970s, tin prices soared to unprecedented heights, hotly pursued by the cost of production. If the price of tin fell, the whole Cornish mining industry could shut down almost overnight. Even in four years, South Crofty had been through several cycles, from hiring every able-bodied man they could find to pinning up layoff notices in the dry. John was not about to anchor his whole life to this situation, when the limitless opportunities of Canada were already open to him. Nevertheless, it was with a heavy heart that John said farewell to his friends. For him, the heat of high summer would always be tinged with the sorrow of that parting.

12

WINTER

On Friday the thirteenth of August 1976, John left the land of his birth and childhood and went to Canada. South Porcupine was unchanged since he had left it two years before. The dusty main street was quiet in the midmorning heat. At the far end of Porcupine Lake only a narrow band of trees, speckled with houses, separated the sky from its reflection in the still water of the lake. The low, flat horizon, broken by the silver-painted headframe of the Pamour mine, seemed to limit John's view of his future in the same way that it limited what he could see of the physical world about him. He sought to dismiss from his mind the vague disquiet that this caused.

John thumbed a ride out to the mine and went to the office to report his arrival to Bob Perry. The general superintendent was not in an optimistic mood. The price of gold had just fallen abruptly to $100 an ounce from the $180, or even $200, it had reached in the preceding six years. It cost the Dome mine $105 to produce an ounce of gold in those days, so this turn of events was viewed with alarm. Bob Perry looked pensively out of the window of the little dark room with green linoleum on the floor that served him as an office.

"I don't know," Bob remarked slowly. "The way things are going, we might all be looking for a job. We can't offer you anything definite, but we'll hire you on anyway and see what happens." Most companies in that position would have sent John on his way, but Bob Perry had offered him a job a year before, and Bob was as good as his word.

John was to report to Bob Drynan on night shift, to work with Marcel and Smoky in 1194 stope. He went back to South Porcu-

pine and lay on his bed in the rooming house, looking at the ceiling.

John had found accommodation with more amenities than the Algoma House Hotel offered; now at least he could make breakfast and fill his lunchpail before going to work in the morning. One of the better-known rooming houses in South Porcupine was run by a Mrs. Toschak.

Mrs. Toschak was Ukrainian and had come from Saskatchewan in the depression. She was nearly sixty, 5'6", trim, with a mouth that smiled and china blue eyes that did not. Her husband, Jake, had been born and raised in northern Ontario and was a mild, good-natured man. Jake's current project at that time was the resurrection of a black-enameled typewriter dating from the 1930s that he had found in the city dump.

Other roomers besides John occupied the other rented rooms at Mrs. Toschak's, mostly students at the nearby Northern College. They worked different hours than John did and were distinguished in his mind chiefly by the violence or otherwise of their final departure. Besides running a rooming house, Mrs. Toschak sold tombstones that came from a quarry near Cochrane, sixty miles away.

It was already the middle of August when John returned to South Porcupine. After a week or two of summer heat, nocturnal temperatures began to plummet, and the September nights were frosty. The first snowflakes of autumn fell with the dead leaves. Frost followed frost, snow followed snow, as the sun's daily apogee sank and its heat weakened in the sky. John had mailed his thick duffel coat to himself, packed in a cardboard box. Sent from England in the summer heat, it arrived in South Porcupine just in time to ward off the winter cold, which came fierce and early that year.

Each morning on the day shift John tumbled out of bed in disgusted disbelief at the demand of the alarm clock. The light from the bare bulb stabbed his sleep-drugged eyes. His muttered obscenities hung on the hot, dead air of the room. He crept downstairs, avoiding the two steps that creaked, quietly made his breakfast and filled his lunchpail. It took him a full five minutes to wrap himself against the cold before going out into the quiet, dark street. Snow muffled the town in a white shroud of silence.

John joined the silent crowd at the bus stop, greeted by shuffling and grunts of recognition. Their breath hung about them in a cloud of vapor. The bus arrived. With an orderly motion, the men flowed on board, their fares clinking in the cash box, dispersing into their private selves as they sat down. The bus driver was new to the job; as they left the lights of town behind them someone prompted him, "Hey, partner, why dontcha turn that #$%*& light off?" The bus went dark, filled with shapeless forms brooding wistfully on their interrupted sleep. Even the first grayness of dawn was still an hour away. The headlights lit the snowy road, while the dark bush blurred by on either side.

The bus pulled into the lighted mine yard. The men flowed out, merging with the stream of those arriving by other means. In the dry, by John's locker, there was always the same man; John never knew who he was or what he did, but he was always there at the same time, sitting, smoking. "Well, young fella, another day, another dollar." John winced at the repetition of this inane cliché and returned some equally meaningless rejoinder. He took off his street clothes and padded past the time clocks, which had been faithfully recording the passage of each shift since they were made in 1929 by a company called International Business Machines. He dressed in his thick, scratchy, dirt-stiffened mine clothes, putting on his heavy, ribbed mine boots, hard hat, and lamp belt. And so the shift began, as it did for maybe 30,000 other men in the mines of northern Ontario.

To be sure, John had worked in more than one mine, but it had always been for a few months before moving on to advance the main course of his career. The idea that a succession of mornings like this, stretching away down the years, might actually be the main course of his career had not occurred to him. John had much to learn at Dome; career prospects within the company should have been amply satisfactory to someone in his position. The fall in the price of gold was short-lived, and optimism returned to the mine. Nevertheless, a question floated through his mind.

While at mining school, John had looked forward to the time when he would no longer have to study every weekday evening. The relief from mental pressures was enjoyable, and John made good use of his newfound freedom to visit the various watering holes around South Porcupine. Some bars were divided into a section for "Men Only" and another for "Ladies and Escorts," differentiated by the presence or absence of carpet on the floor. The intention was to protect the fair sex from the rougher elements, but some women drank harder and swore worse than men, so it made little difference. They say that in a bar in a Canadian mining town the sound of drilling is so loud that you cannot hear, and the smoke from blasting so thick that you cannot see. Figuratively this is true, for the mine is an unending subject of conversation.

Three elderly men sat at a table. John overheard the same monologue, repeated again and again in various forms, with great emphasis, arm waving, and blasphemy. The speaker had been on one of the first shaft-sinking crews in the Blind River country in the early 1950s. Back in those days, there had been no shaft jumbos or Crydermans, and they sank the shaft with "them @#$% CR58 pluggers" and shoveled the broken rock by hand from the shaft bottom after each blast.

That winter had been a cold one, with temperatures of fifty degrees below zero for weeks on end. The shaft and the bunkhouse and the cookhouse formed the sum total of the men's lives out there in the deeply frozen bush. Every two weeks the crew were trucked into Sudbury to be bathed and deloused. The bunkhouse became so infested with bedbugs that they burned it to the ground at the end of the job. When the shaft was complete, the speaker had moved on to the next development job, rather than accept the lower pay and comparative monotony that would follow when the company crews took over.

Some people might think that a shaft sinker would like nothing better than to stay on at a working mine after its construction is complete, to enjoy the more orderly existence and greater amenities. They misjudge the tramp miner. When the production company takes over from the development contractor, so does a different breed of miner and manager. For the tramp miner, the frenetic pace of highball mine development is his fix; the knowledge that only the smart and the tough can survive and earn the big bucks is his validation of himself. So he packs his gear and moves on to the next job for Paddy Harrison, MacIsaac, Redpath, Cementation, TMCC, or any other of the contractors who have come and gone, of which only the best have survived.

A little man stood up, short but solidly built, dressed in a navy-blue trench coat, which gave him an inexplicable resemblance to a Polish admiral. He announced to the assembled gathering in a rasping voice, regardless of company, "I can go to them tomorrow and tell them to stick their (expletive) job up their (expletive) ass. There's guys like Paddy Harrison phoning

me every (expletive) week wanting me to go to work for them.
'Cos, partner, there ain't too (expletive) much you can teach me
about this (expletive) racket—shaft, drift, raise, the whole (expletive) issue. An' you wanna (expletive) remember, partner, that
this here" (jabbing his chest with his thumb), "is The Fox, one
of the big names in northern Ontario mining." With that, he
stalked out into the night.

Such is the tramp miner's boast. The company knows that it
is true. The hometown miner knows that it is true, too, and it
galls him because he fears secretly that he could not live up to
the demands of a mining contractor, or even of a different mine.
He can never make that boast and prove it. So he snarls and
whimpers and attends union meetings and wishes that he, too,
could tell the company what to do with their job. But when all
is said and done, he lacks the self-confidence to go out and
prove it. Besides, he has a wife and children to care for, and a
mortgage and other debts to pay off, and he cannot risk taking
that plunge out into the big, wide world. These ties will keep
him a company miner until he retires, and the company gives
him a retirement party and a cheap watch and, his mainspring
broken, he lies down and dies.

There was the scene, all too often repeated, of a small group
drinking together as friendly as could be. Then the beer would
start talking, the self-defensive bragging would well up, and
tempers would be on edge. The explosion would follow—an
insult, a splash of beer thrown in someone's face, a scraping of
chairs, a tinkle of glass breaking on the tiled floor, futile re-
monstrances. The other patrons would fall silent and turn in
their chairs. After some preliminary pulling, hauling, and name-
calling, the antagonists would adjourn to settle their differences
outside.

On one occasion the antagonists were both women. One of
them, tall and solidly built, stood up and yelled, "!@#$% you,
Darlene, you @#$%&* bitch! Just because I wear a red dress
and I come from Sudbury, you think I'm a @#$%& whore!" She
fell to the floor with a crash; the party broke up in disorder.

In the Porcupine, as in most towns in northern Canada, drink
was (and is) a demon. Jobs, careers, marriages, and small businesses went down like ninepins as all sense of time and responsibility were blown away by alcohol and by the camaraderie of
the mine and the beer parlor. It was easy, so easy, to head for
the bar after the day shift.

"Come on, Joe's buying!"

"Yeah, see the moths fly out of his wallet, the cheap bastard!"

"OK, just a couple, though, I got to get on home."

The beer would start flowing, and the drilling and blasting would get going. In the bar it was warm, while outside the temperature was nudging zero, and snowflakes swirled through the light of the street lamp. Besides, what did a guy have to go home to anyway, and—what the hell—it's nine o'clock already; sure, let's have another jug. And so a man would stagger home at one o'clock in the morning, flop into bed in a stupor, go out like a light, and then drag himself out to the mine at half past five, because if he was man enough to get drunk at night, he was man enough to go to work the next day. Getting drunk was considered a normal occupation of spare time; some kept the demon under control or fought it off; others were powerless in its grip.

Mrs. Toschak kept the house heated like an oven. Hot air and the smell of stale cooking collected on the top floor. Coming back to his room at night, John would peel off several layers of clothing in a hurry, but even so, he would break into a sticky sweat. One evening, desperate to let a breath of fresh air into the room, he tried to open the window. Mrs. Toschak had nailed it shut. That was how it stayed. As the winter wore on, the coating of ice on the glass grew thicker, and it became impossible to see out.

One Saturday, John went back to his lodgings after eating lunch at Jim's Restaurant. He crept upstairs and lay on the bed, looking at the ceiling. He decided that since it was not too cold, a walk would do him good; snowflakes floated lightly through the quiet air.

John walked down the street to the railroad station. It was deserted. Nothing came through on Saturdays. John looked about him and turned to follow the track, which lay on an embankment just above the level of the swamp. South Porcupine was soon lost to view in the falling snow.

Where the railroad track crossed the back road to Timmins, one of the immense Dome tailings dams stood in the angle of the road and the railroad. John scrambled up the slope of soft, muddy sand.

The crest of the dam was the highest point of solid ground for some distance. The tailings pond impounded within the rectangular dam was a mile across, its level surface broken only by a clump of dead trees. Clouds of snow blew across it, driven by a fitful wind. A patch of gray marked the half-frozen water that gathered about the central decantation weir. On the far side was the dragline, which worked its way forever around the periphery, building the dam ever higher above the surrounding bush. At the other end of the dam the Dome headframe stood among the trees. Two miles away in the gloom was the headframe of the Paymaster no. 5 shaft.

Back in the thirties and forties, the Paymaster had been a hive of activity, compressors thudding, lights spangling the darkness of the empty bush at night, the hoist running, the rock skips dumping one after another, shift crews coming and going, feats of drilling and mucking accomplished far underground to be recounted and argued over in the bars of South Porcupine. Now only the towering headframe remained, its torn corrugated iron flapping and creaking in the wind, glass snapping underfoot, the ground littered with scraps of cable, rusty nuts and bolts, shattered timber, and pieces of bent pipe. An overturned muck car lay half hidden in the encroaching bush.

Joe Raymond was a shiftboss at the Dome, where he had worked for years. His brother, Peewee, never worked anywhere for years; "months" was more his style because he was a tramp miner. Peewee had worked at the Paymaster in 1948. He missed the bus one morning and was late for work. That was the morning the hoist cable broke and the cage, loaded with men, fell to the bottom of the shaft. Peewee would have been on that trip.

John turned about and slithered down the face of the tailings dam. Following the road toward the lights of South Porcupine, he thought of the steak he would have for supper at Jim's Restaurant.

Affairs at Jim's were not always as peaceful as the industrious Jim and his wife would have liked. One evening, John went there for his meal before going to work on the night shift. Tables and padded benches were arranged around the walls. In the middle of the room were two rows of tables and benches separated by a low wall, shoulder-high to seated guests. John went to sit by the wall farthest from the door. The only other patrons were Wally and the Woman, who were sitting in the middle of the room. John knew them both by sight.

The Mighty Mouse came in, well fueled, to buy cigarettes. The Mighty Mouse was a bandy-legged little man with a rasping voice and a grudge against the whole world. It was the

Mighty Mouse who started barroom fights by picking on a man bigger than himself. It was the Mighty Mouse who stood up at union meetings and denounced the committee as scabs in the pay of management. It was the Mighty Mouse who, sitting on his upended lunchpail, waiting for the cage, applied a cigarette lighter to the crotch of the Toad's oilskins. Something happened very quickly, and there was the Mighty Mouse looking sorry for himself with blood running from his nose; the Toad stood impassive as before. The Mighty Mouse was a hazard on two bandy legs.

As was inevitable, a shouting match broke out between Wally and the Mighty Mouse. John heard the Mighty Mouse call the Woman a "!@$% rat." Wally sprang up, grabbed him in a half-nelson lock, and frog-marched him out through the door, the Mighty Mouse shrieking abuse in both of Canada's official languages. Wally came back, but the peace was short-lived. After a commotion outside, the Mighty Mouse came back in as though jet-propelled. John heard the shouting break out afresh but took no notice.

John heard a bang, and something struck him a glancing blow on the scalp. He and his meal were deluged with broken glass. The Mighty Mouse had seized a glass sugar-shaker and thrown it at Wally. The shaker had struck the low dividing wall in the middle of the room and shattered into a thousand fragments; John had been directly in the line of fire. Wally went for the Mighty Mouse. The glass door survived by a miracle. Noises off suggested a full-scale, knock-down-and-drag-out fight in progress outside, until flashing lights and the stern voices of authority broke it up. John picked the glass from his meal, ate the remainder, chewing carefully, and went to work. Next day, Jim presented him with a steak supper by way of apology. Jim was like that. The day he won a lottery, Jim gave each one of his regular customers a free meal.

As autumn wore on, the cold deepened; the snow deepened with it. John came back from work just after dawn one morning to see the thermometer on the outside of the house reading forty-four degrees below zero. The snow on top of the lake ice, brushed by the wind into swells and ridges and ripples, was tinged pink by the rising sun. The lake was frozen hard enough to drive a truck on the ice. Each sound rang in the bitter air, whether of human activities or of a tree cracking in the deeply frozen bush. Metal would burn exposed skin, and the cold gnawed at the tips of ears and noses. Moisture froze in the nostrils with each indrawn breath.

John caught a cold. He ignored it; a man could not miss work every time he had a runny nose. The cold turned into a fever and a barking cough. He reckoned that as long as he was well enough to stand upright, he was well enough to go to work. Sick roomers had no place in an efficiently run rooming house. Besides, he had to go to the restaurant to eat, and sitting indoors all day would be worse than going to the mine, where there was plenty to occupy his mind, and the damp air was easier on his sinuses. The air in the rooming house was so desiccated that John slept with a damp rag over his face.

John noticed that his chest hurt, and he could not draw a full breath. The doctor said he had pleurisy, but only in one lung, and gave him some pills. The pleurisy scarred his lungs, and for the rest of his life, whenever he laughed deeply—and the occasions were numerous—he would choke, and in so doing he would remember that winter in the Porcupine. Other men were at work in the mine far sicker than he, and with worse things wrong with them. If they drank too much because it made them feel better for a while, who would blame them?

Each evening, John ate supper at Jim's Restaurant. The food was inexpensive, wholesome, and bland. One song on the radio seemed to be an enduring favorite—"The Wreck of the Edmund Fitzgerald." The tune ran on, monotonous, mournful, and subdued, never quite repeating itself. It was like northern Ontario. John wanted to throw his plate at the radio, but he could not escape the song and its dreary sadness. The song ended; John paid the bill; he slithered and stumbled through the snow back to his lodgings. He crept in, took his boots off, and moused his way up to the little, overheated room where the furniture creaked at him and the smell of stale cooking caught in his throat, secure in the promise of another restless night cut short by the demands of the mine.

Through much of that winter John functioned like an automaton, too sick to do anything except work, eat, and sleep— too sick, indeed, to pack up and go back to Cornwall. Mrs. Toschak eyed him thoughtfully and one day unburdened herself of a single sentence: "You are a very sick young man I give you a real good deal on a tombstone Air Canada will take it back to England I know someone who will make all the arrangements."

One evening, on his return from Jim's, John threw his hat, coat, and gloves onto the bed and sat on it, looking at the floor. He meant to do something, such as write a letter, read, drink a tepid bottle of beer from the case under the bed, or perhaps go out to the bar. Instead, he sat, undecided, and stared at the floor. After pondering these options for half an hour, he swung his feet onto the bed and lay staring at the ceiling to gain a better perspective on the problem. In a sudden fit of determination, he swung his feet off the bed and sat up, intent on doing something constructive, but compromised by staring at the floor. Gradually, he devoted more attention to the polished floorboards and threadbare rug, and less to deciding what to do. After a while, a noise disturbed him, and he looked at the clock. Noting, without surprise, that he had been staring at the floor for three hours on end, he undressed and went to bed.

The next evening, it was getting dark as he made his way to Jim's past the big, three-story rooming house known as the Finn House. The Finn House was a standby for some of the Dome miners because, unless a man made it for himself, it was the only place where he could get a square breakfast before going on the day shift. It was, however, always changing hands. There seemed to be an unending succession of broken windows, fights, and people being thrown out. Losers stayed there. One tenant tried to shoot himself, but he could not even succeed at that, and they carted him off with his face blown apart.

The windows of the basement rooms of the Finn House extended a foot or two above ground level. The troglodyte tenant could, therefore, choose between privacy and letting in whatever daylight there might be. John happened to look down into one of the lighted windows to see a middle-aged man, dressed only in the coarse, gray long johns that everyone wore at the mine. He was sitting on the bed, staring at the floor. Staring at the floor was a popular pastime in South Porcupine.

The interminable winter dragged on through January and February, and into March. Much to Mrs. Toschak's disgust, John found himself an apartment. He bought some furniture and appliances, thereby adding a new dimension of permanence to his life. At the same time, he went to work in evenings and on weekends at Bruce Watters's printing press. Bruce was a mill foreman at the Texasgulf concentrator; in seven years he had developed such a thriving part-time business with a printing press in his basement that he was having to turn away work.

John had picked up the skills of letterpress work at school. Part-time typesetters and press operators do not grow on every tree, so Bruce's basement became a haven of warmth in winter and coolness in summer, not to mention the lavish hospitality that Bruce and Pauline showered upon him, and a wage that helped to pay the rent.

Not many immigrants will forget their first winter in northern Canada; the experience is unlikely to be recalled with pleasure. But not even the northern Ontario winter lasts forever. In March came two heavy falls of snow, but the air lost its bite. In April liquid water began to appear out-of-doors. One afternoon in May, John came out of the mine to find that it was summer.

The air had changed during the night. John had felt it that morning as he waited for the bus. It had been mild and damp, as if the icy casket of winter had been broken open. That afternoon it was warm. Miners stood about the gate waiting for their rides or buses, jackets slung over their shoulders, smiling in the warmth of the sun, squinting at its light. They looked up into the blue sky, soft with the promise of summer days to come, while for so long they had stared at the ground, hunched against the cold, cowering under the leaden skies of winter.

All winter long the air lay on the ground, cold and sluggish as if made of transparent iron. When it moved, it was vicious, knifing through clothing, slashing at faces, biting the tips of ears, noses, and chins. The lungs dragged it in through the unwilling nose and throat to extract oxygen and pushed it out again in a cloud of vapor. In a matter of days the air had changed. It begged to be breathed; like wine, it was there to be enjoyed. It did not yet carry the fragrance of the newly awakened bush, but only because it was an invitation freshly issued, a canvas waiting for the painter's brush, the silence of a concert hall awaiting the imprint of the opening notes.

Rivulets of water formed in the streets. Where no liquid water had been seen since November, now it chuckled to itself as it streamed, sparkling, from the shrinking snowbanks. The packed snow on the roads melted, leaving riffles of the grit that had been scattered on it. Here was a garbage pail missing for months, a child's glove, a shoe, a beer bottle—all enfolded in the piles of snow that were now vanishing under the growing strength of the sun. The lake was rising under the melting ice, gray and black with slush.

It was quitting season at the mine. Men scattered in all directions, to contracting jobs, to construction, to work in the bush or indeed, out of sheer restlessness, to other mines. Quitting season came around every year; few were immune to its call.

John recovered his health, and to some it might have seemed that he was set to become a "lifer" at the Mighty Dome.

13

THE PORCUPINE AND SUCHLIKE PLACES

"The Porcupine camp," or "The Porcupine," was an all-embracing term describing seventy square miles of forest and lakes, muskegs, and slow-flowing creeks, whose level topography was in places interrupted by scarcely perceptible hummocks and ridges of bedrock. A rockpile to the west of Timmins, barely 300 feet high, was named Kamiskotia Mountain, while a part of South Porcupine south of the railroad tracks, rising perhaps 30 or 40 feet, was known as Connaught Hill.

The camp contained four main settlements and totaled some 40,000 people. Biggest was the city of Timmins. A little to the east was the settlement of Schumacher. Six miles to the east through the bush was South Porcupine, commonly known as Southend, from its position at the south end of Porcupine Lake. At the north end of the lake, and within an hour's brisk walk of South Porcupine, was Porcupine. Porcupine had two subdivisions—Golden City and Pottsville. A few small settlements were scattered along the farmland of the Little Clay Belt, 30 miles farther east, but the nearest settlement of comparable size was Sudbury, 200 miles away. The towns and villages of the camp existed because of the mines.

Following the discovery of gold in 1908, the gold mines grew in number over successive decades until, in the late 1930s, roughly two dozen were in production, ranging from the Hollinger, McIntyre-Porcupine, and Dome, whose names were well known to mining people the world over, to mines like the Faymar, Moneta, and De Santis, which were almost unknown. By

the 1970s only five gold mines remained at work in the camp, the rest having closed because of rising costs and a fixed gold price, exhaustion of ore, or some combination of the two.

The biggest of the gold mines had been the Hollinger, on the outskirts of Timmins, which ground out 8,000 tons per day, decade after decade, until the ore was exhausted and the mine was abandoned in 1968. By that time its deepest levels were a mile from the surface. Deeper yet, and still active in the 1970s, was the McIntyre-Porcupine, across Pearl Lake from Schumacher, which was the deepest mine in the camp. A succession of internal shafts went down more than 8,000 feet. Five miles to the east of South Porcupine was the Pamour, whose silver headframe could be seen from the shore of Porcupine Lake. A short distance from the back road to Timmins was the Aunor-Delnite, a dead-and-alive thing perpetually on the brink of closure.

Besides the Dome, of which we will hear more, these were the gold mines of the Porcupine camp. In the 1970s Noranda Mines Ltd. owned all of them except the Dome, which was owned by Dome Mines Ltd., an independent company. Dome Mines Ltd. also owned the Sigma mine at Val D'Or, Quebec, and the Campbell Red Lake mine at Red Lake, Ontario, the richest gold mine in Canada.

Ten miles from South Porcupine, down a sandy, gravelly road, was a small nickel mine named the Langmuir, also owned by Noranda, while other mines were worked for nickel and copper at various times in the remoter environs of the camp.

Twenty miles north of Timmins was the enormous Kidd Creek mine, which extracted 15,000 tons per day from an immense body of massive sulfide ore containing copper, lead, zinc, silver, and several other metals. Expert geologic opinion had at one time stated that an ore body of that nature could not exist in that type of rock, yet a geophysical anomaly had been known for years, so large and so pronounced that the International Nickel Company of Canada Ltd. flew over it regularly to calibrate their airborne geophysics instruments.

Then a two-bit company by the name of Texasgulf Sulfur Inc., whose sole claim to fame was a Frasch sulfur operation down on the Gulf coast, diamond-drilled the anomaly in the early 1960s. The result was a mine that employed more people than all the gold mines combined. If the tin contained in minute amounts in the Kidd Creek ore could have been extracted economically, that mine would have produced more tin than all the tin mines of Cornwall.

With such a dense concentration of a dangerous industry such as mining, much emphasis is placed on preparedness for first aid and rescue. Training is encouraged by annual competitions, which bring out a keen competitive spirit between one mine and another. The annual mine rescue competitions are a bigger event than the first aid competitions. Part of the reason is that any corner store can field a first aid team, but only the mines have rescue teams. Possibly the function for which the mine rescue teams are trained lends an aura to the competitions.

Ontario has a highly developed mine rescue system operating under a provincially sponsored organization known as Ontario Mine Rescue. The organization was founded in 1929, the year after a disastrous fire underground in the Hollinger mine. Alf Stanlake, one of John's Cornish friends in South Porcupine, was brought out of the mine on a stretcher, believed dead. Fortunately he was resuscitated with no lasting ill effects and lived another fifty-five healthy years.

Fire arouses a deeper dread than most other hazards of underground mining. Various things can catch fire in a mine; dry timber, oil, grease, and garbage are the most obvious, but other things can also burn. Once a fire is started in, say, an accumulation of grease on a diesel engine, other substances, such as tires or track ties, can ignite and become fuel in their turn. The danger lies not so much in the fire itself as in the smoke and poisonous gases that fill the mine and in the consumption of oxygen by the fire.

Over the centuries thousands of miners have been killed by mine fires in numbers ranging from ones and twos to hundreds at a time. The death toll continues year by year. The Sunshine mine disaster of 1972, in which ninety-one miners died, was a mine fire believed to have started by spontaneous combustion in old timbers. The news reached John only days before he went to work at South Crofty.

The function of a mine rescue team is to go down the mine after a fire has been detected, find and rescue any survivors, and then find and extinguish the fire. The team may have to go underground, wearing oxygen breathing apparatus, into an intensely poisonous atmosphere in which the darkness of the mine is so compounded by smoke that vision is impossible.

These conditions may be aggravated by intense heat caused by the fire itself.

It may occur to the gentle reader that deliberately entering a burning mine and stumbling blindly through a maze of smoke-filled workings in ovenlike heat, depending on a limited supply of oxygen obtained through a mask clamped over one's sweating face, contains the ingredients of the type of nightmare that most people hope never to have. Yet such is the eventuality for which the six-man rescue teams are trained. It is not a task for those of a nervous disposition, as a frightened man breathes faster and so accelerates the depletion of the supply of oxygen on which his life may depend, as well as shortening the time available to the team as a whole, whose return to fresh air is dictated by the man with the least oxygen remaining. In the event of a serious fire, rescue and fire fighting may go on for weeks, with relays of teams working around the clock, rather than within the two-hour limit of a team's breathing apparatus.

Mine rescue men must be as rigorously trained as fire fighters, perhaps more so because of the extreme horror and hazard of a mine fire. A team becomes a close-knit unit from which any man showing signs of a nervous, unstable, or undisciplined temperament is ejected without a trace of compunction in case he should endanger the team in a real emergency. Regular monthly training is interspersed, twice a year, with full-scale call-outs, during which the mine is evacuated, and the rescue teams do not know, until they are fully equipped and ready to go underground, whether it is another practice or the real thing. Anyone entitled to wear an "Ontario Mine Rescue" sticker on his hard hat does so with pride.

Each year a provincewide competition is held in which every mine worthy of the name competes. First, each mine fields a team in a local competition. The winner of the local goes on to the provincial, which is held in a different mining town each year.

Naturally, every mine rescue man in Ontario longs to be on the team hailed as the best in the province, but the Porcupine mines had to realize that the winner of their local would be matched against big-time competitors. In particular, the mines at Elliot Lake were known to allocate more paid man-hours to mine rescue training than the Porcupine mines were willing to do; as a result, the teams from Elliot Lake put on an extraordinarily polished performance that was a privilege to watch.

"Watch" is perhaps a misnomer, as a mine rescue competition has close to zero value as a spectator sport. The competitions tend to be carried out in dark indoor arenas in a maze of burlap walls that simulate the drifts of a mine. The metallic clinking of equipment, shuffling feet, wavering mine lamps, and words of command muffled by the face masks of oxygen-breathing apparatus are all that reveal the activities of the team "on the floor." A trained mine rescue man can, nevertheless, discern from those faint signals not only what is going on but also how well a team is progressing.

John trained as a mine rescue man, but in four years in the Porcupine he never had to fight a mine fire, which fact he regarded with the attitude of a soldier trained for a war that ends before he is blooded in battle—a frustration mingled with relief. No amount of training and competitions could ever tell how a team would fare in the real thing.

One time, however, a persistent smell of smoke was reported on the Dome 2,000-foot level. It was feared that a smoldering fire might be ready to burst out, so the next shift was told to stay home. John and another man were sent underground with gas masks to reconnoiter. They found nothing to report except for that faint smell of smoke. Several shifts' work was lost before the smoke was traced to the top of no. 5 winze, a small internal shaft between the 2,000- and 2,300-foot levels.

The winze had been sunk in the 1920s to explore ore discovered by diamond drilling below the deepest level that existed at that time. By the late 1970s it had so far outlived its usefulness that it was, to all intents and purposes, disused. Above the 2,000-foot level at the winze was a cubbyhole cut out of the rock for some forgotten purpose, so difficult of access that even its existence was almost unknown. That was why one of the samplers had chosen it as a place to sleep during the latter half of the shift.

The samplers were the men who chipped rock samples from the faces of development headings and took them to be assayed. Sampling done properly is hard work, but at Dome it was seldom done properly, and the "sample gang" was regarded as having a soft job. It was good that such jobs existed, as men could be assigned to them who were debilitated by illness, injury, or too many years of hard work in the mine. Some such jobs, however, were occupied by men who were able-bodied but lazy, and the sample gang was notoriously idle. This particular man had built himself a nest of burlap in a place where no one would even think to look for him. Awaking suddenly and finding that he was late for the cage to the surface, he had left a lighted cigarette end in the burlap, which smoldered and infected the vicinity with a smell of smoke.

The culprit eventually owned up and was fired. John happened to be in the pay office one day and overheard one of the clerks remark that dismissal seemed too harsh a penalty. Considering the disruption caused by the incident and the lethal danger of mine fires in general, John burst out, "Too severe? They should have jailed the bastard!"

One crime, specific to the mining of gold and precious stones, could land a man in jail, so it was said. That was "high-grading." Gold is one of the few metals whose occurrence is such that small chunks of high-grade ore may be valuable enough, and easily enough processed, to be removed from the mine and sold for sums of money that are significant in terms of a day's pay. Not all gold mines contain ore rich enough to be worth high-grading. Whether high-grading ranks as a sin, crime, misdemeanor, or perquisite depends on point of view. The miner regards it as a perquisite; the mining company regards it as a crime. Consequently, "to high-grade" is a euphemism, uniquely contributed by the gold-mining industry, for that good old English verb "to steal."

Dome, and most of the other Porcupine mines, contained high-grade in substantial quantities. High-grade ore was richer and more plentiful in the Dome's earlier years, but even so, in the 1970s visible gold was of common occurrence, and real jewelry rock far from rare. One legend, never confirmed, denied, or disproved, concerned the existence of stopes so rich that they had been sealed off against the day when their ore would be needed to save the mine from closure. No one had ever worked in such a stope, nor could anyone say where one might be, but the legend persisted.

No shadow of doubt clouded the minds of the mining companies: high-grading was theft. Even the possession of high-grade was said to be a crime, although John never looked in the law books to see if that was true. It was said that half the small businesses in Timmins had originally been financed with high-grade liberated from the Hollinger mine. When the old Dome dry burned down in 1929, the new version was intended to be a model of its kind, specially designed to prevent high-grading.

On entering the dry at the start of the shift, a man took off his street clothes and hung them in a locker. He walked down a flight of steps and passed the time clocks to where his mining clothes hung on a grappling hook hoisted up to the ceiling. Both parts of the Dome dry were cavernous and dark. The mine side was darkened additionally by a black deposit of aluminum dust that was blown into the air before the arrival of each shift crew as a preventive against silicosis.

Silicosis was a dark secret of the Porcupine, about which more was rumored than known. The disease is an injury to the lungs caused by inhaling microscopic particles of silica-rich rock dust, which is generated by drilling and blasting and is stirred up by mucking or chute-pulling. Tiny, sharp particles scar the lungs irreversibly, progressively impairing their functioning. If the exposure to silica dust is sufficiently prolonged and severe, death results, in a particularly unpleasant form. In the Porcupine, men in their sixties often died of pulmonary complaints—lung cancer, pneumonia with or without complications, "emphysema"—but silicosis was seldom diagnosed as the primary cause of disablement or death, and few men (or their widows) collected silicosis pensions. "Aluminum prophylaxis" was thought to reduce the scarring effect, but aluminum dust may also have masked the lungs on the chest X-rays, which are a prime diagnostic for silicosis.

In any case, mining companies in the Porcupine believed that aluminum dust helped to prevent silicosis and blew it into the air of the dry before every shift crew came to work. The dust settled on everything and helped to make the ill-lit dry a place of stygian gloom.

After coming up from underground at the end of the shift, each miner handed in his lunchpail to be searched, stripped off his wet, dirty mining clothes, threw them in a heap on the floor, and headed for the communal shower room with shower heads all around the tiled walls. Having showered, he dried himself in a well-heated and similarly communal drying room. He hung his mining clothes, towel, and washcloth on his grappling hook, pulled it up to the ceiling by means of a chain running over a pulley block, and secured it with a hook. He walked past the time clocks once more and upstairs to the street clothes lockers.

Mine security guards patrolled these areas; one of them had the unenviable task of standing at the bottom of the stairs as a hundred or so naked men filed past him, watching to see if any-one had concealed a stash of high-grade between his legs or buttocks. Having dressed, each man collected his lunchpail, which had to be marked with his employment number, and headed out of doors for the first time since entering the dry some nine hours before. Towels and laundry were equally sub-ject to search. It was practically impossible to take any object of contraband out of the mine.

The mine staff were not subject to these procedures and were on their honor not to engage in any form of high-grading. They had their own separate dry in an upstairs room. This was not merely privilege for its own sake. In later years John worked as a foreman at a small mine with one dry for miners and staff alike. It was thoroughly bad for the foreman's temper to have all the problems and complaints of the day dumped on him before he even had a chance to change his clothes.

What galled both miners and mine staff was the fact that the mine geologists and samplers collected a steady supply of high-grade to be presented to visiting dignitaries, corporate or otherwise, while visitors of all kinds were, so it seemed, at lib-erty to chip and pick away and take whatever high-grade they could find.

While John was a shiftboss, his integrity was sorely tested one night. Toward the end of the shift he went to visit Rocco, who was working all alone, tramming muck from 1540 ore pass chute. With a flourish, Rocco produced a magnificent specimen of high-grade that had come out of the chute. It was the size of a fist and consisted of pure white quartz intergrown with gold in roughly equal proportions. They both knew that it had to go into a canvas sample bag, and John had to take it to the security guards on the surface. John harbored all manner of unworthy and cynical thoughts about the security guards, not all of which were unfounded, and casually checked with a friend in the ge-ology department that the specimen had, in fact, appeared at its proper destination.

In John's time at the Dome, either there was almost no high-grading, or it was carried out so successfully and in such se-crecy that it was never discovered. The story was told, however, of a man who had put some high-grade into a test tube and had taken the extreme step of concealing it in his rectum. All would have been well but for the fact that in the shower room he was overcome by the urge to break wind. Broken glass and high-grade were scattered all over the floor, and the man was arrested at once and dragged off, stark naked and soaking wet, to whatever fate awaited such felons.

Only one incident came to light while John was working at Dome. Some of the mechanics and electricians worked odd shifts and came and went as they pleased without being searched. In the dry was a small desk, of crude construction and painted green, which contained a logbook of the kind that is maintained at old industrial plants long after its original purpose has been forgotten. Beside the desk was a particularly comfortable wooden armchair which was a popular resting place for anyone whose duties allowed them some brief repose. In the early hours of one morning an electrician coming off shift left his lunchpail on the desk. One of the security guards sat down in the chair, pushing the lunchpail aside with his elbow as he did so. Thinking the lunchpail unusually heavy for the end of a shift, he opened it and found that it was full of high-grade. The electrician was dismissed, but no further action was taken; when last heard of, he had gone to Yellowknife.

High-grade or no high-grade, the Porcupine mines had their own style. From time to time, miners came in from Sudbury or Elliot Lake, but they seldom stayed. Porcupine miners sometimes went to Elliot Lake (to Sudbury almost never), but for every man that stayed there, plenty came back. Miners from elsewhere thought the Porcupine mines absurdly primitive and the wages low, and said so. Porcupine miners retorted that the outsiders knew nothing about real mining, were spoon-fed and overpaid; if they did not like the Porcupine, they were welcome to move on.

In the autumn of 1976, when the price of gold was down and the gold mines looked even shakier than usual, all the enterprising types hit the road for the new Umex mine at Pickle Lake in northwestern Ontario. No sooner had the last men gone than the first were coming back. Accommodation was scarce and expensive and the cost of groceries absurd, all of which canceled out the high wages offered at the mine.

A couple of years later, John learned the technical details. Heading for the shaft one day, he bumped into Scotty, his erstwhile partner at Yellowknife. After the mutual surprise had worn off, Scotty explained that he was working on a geology degree and was one of a party of students visiting the Dome on a field trip. Scotty, the quintessential tramp miner, had, of course, included Umex in his travels.

"Umex" was short for Union Minière Explorations and Mining Co. Ltd. Union Minière mined copper in the Congo and thought that they would try their hand at doing the same thing in Canada. It had not occurred to their Congo-trained engineers that, in northwestern Ontario, the winter weather can be a trifle brisk. Ore from the mine was put through an immense crusher on the surface. Without protection from the weather, the wet muck from the mine froze as it went through the crusher until the thing jammed solid. The crusher was supposed to be handling 4,000 tons per day. Scotty reckoned that, while he was there, it never exceeded a modest 40. Scotty just laughed and moved on. Similar stories that came back from other mines often discouraged experimentation.

As potential employers, a man could choose between mining companies and contractors. Mining companies are those, like Dome, or Preussag at Rammelsberg, or South Crofty, which are in the business of mining a commodity and upgrading the raw product into a form that someone will buy. Their main aim is to produce a steady flow of the commodity at a cost lower than the market price. Their employees are content to keep things running at a steady pace decade after decade.

Contractors are in the business of developing and constructing mines rather than operating them. Mine development is often conducted in places with little or no infrastructure. The job itself is of limited duration—a few years at most—and is totally unstable because everything about it is always changing. The work demands a different mentality to production and places a high premium on resourcefulness, energy, and initiative. It requires men of unquestionable competence who are prepared to work long hours, seven days a week, under the most arduous conditions, and who are content to live out of a suitcase, always moving on from job to job—almost the definition of a tramp miner. Such people can command enormous wages. Between 1950 and 1980, one of the better-known Canadian contractors was a company commonly known as Paddy Harrison.

Morty Mortensen had given it a try working for Paddy Harrison at Shebandowan, a new nickel mine in the bush west of Thunder Bay. Morty had been working in the Dome's highball stope—2415. Twenty-four-fifteen produced a steady 7,000 tons a month. A good crew worked there, and they liked it that way so as to keep the big bonus paychecks coming—farmers and lunatics need not apply. One day, the shiftboss or stope boss rubbed Morty up the wrong way, so he collected his time and made tracks for Shebandowan, which was where the action happened to be in those days.

Paddy Harrison was driving a decline, among other things, so they gave Morty a jackleg and a pair of hip waders, pointed him in the direction of the face, and told him to have at it. Twenty-four-fifteen was warm and dry, so Morty found it difficult to summon up any enthusiasm for a place where he had to work in hip waders, especially where the water was as cold as it was at Shebandowan. He was obliged to be philosophical, however, by the fact that Shebandowan was a thousand miles from home.

He was less inclined toward philosophy when some nasty little man insulted him in the camp that evening. Morty naturally pounded the daylights out of him and thought nothing more of it.

The next morning Morty went across to the cookhouse for breakfast. A few of the feeding men looked up at him dully, but he recognized no one. He took his tray of food from the serving counter and found a vacant space at a table. Someone approached the table. Morty looked up. It was his antagonist of the night before. They recognized each other simultaneously, just in time for Morty to duck beneath the tray and its load of food, cutlery, and dishes, which came flying at his head.

The cookhouse erupted into a pandemonium of fighting men. Furniture, food, crockery, and cutlery flew in all directions. The windows were gone in no time. The mingled uproar filled the room until it seemed the walls must burst. Morty, having dealt with his opponent, managed to escape through the side door in time to see the security guards rushing the front door with pick handles.

Picking his teeth and flicking food from his clothing, Morty headed for the portal. The superintendent met him on the road and, with a small circling gesture of his index finger, directed Morty to the time office. A week later, Morty was back in 2415 stope.

The Kid had been born and raised in the Porcupine. His father worked at the Dome, and so, eventually, did the Kid. By and by, the Kid began to think himself a pretty fair miner and reckoned he would give it a try in the uranium mines at Elliot

Lake. Maybe it was springtime and the sap was rising, or he started hearing stories of big paychecks at Elliot Lake and listening to them, or his stope made a poor bonus one month, or perhaps all of the above. So he left the Dome and went to Elliot Lake.

The uranium mines at Elliot Lake had had a checkered history. Uranium was discovered in the area in the early 1950s. Toward the end of the decade, with big supply contracts across the border, the Elliot Lake camp, which had been wilderness in 1950, was booming flat out. Underground development went on at a furious pace, with little thought for safety or dust control. In those days everyone smoked. It was not until fifteen or twenty years later that men who had worked at Elliot Lake began to die of lung cancer in suspiciously large numbers. Only then was the deadly combination of uranium-ore dust, cigarette smoke, and the eerily named "radon daughters" identified, but by then it was far too late. In 1961 the Americans canceled their supply contracts, and Elliot Lake became an instant ghost town. Anyone who had bought a house lost their money. During the boom years, so many people worked at Elliot Lake at one time or another that John would meet dozens of them. Never once did he meet anyone who remembered anyone else.

In the middle 1970s, the camp burst into life again. Again there were big supply contracts, this time with the Ontario Hydro and Power Authority. The mines were reopened, and again the pressure was on. Mines (plural) cranking out 10,000 or 15,000 tons per day were required to materialize out of thin air yesterday, and preferably sooner. So grueling was the pace of the work underground that a shiftboss working for a contractor would burn out in a year and would have to take his earnings and find something easier on his system. Men lived in tents until they could get a room in a bunkhouse; the price of most commodities made like a Saturn Five; and Stompin' Tom Connors belted out his incomparable hardrock mining song "The Shores of Elliot Lake."

The attitude of the Dome miners to their employer in those days was summed up in various sayings that were current around the mine. "They don't pay much, but they sure treat you good." If a man wanted to go away and break his neck for a few extra dollars, let him. "Nobody bothers you." The shiftbosses were not on the men's backs all the time, like they were said to be at Sudbury. "What doesn't get done today gets done tomorrow." The ore had been there for a billion years, so a day or two extra did not matter. The mine kept on producing 2,800 tons per day, five days a week, and everything went along fine and dandy. That was why the Dome was a happy outfit where as many as three generations of the same family might be working at the same time. The Kid found that the mines at Elliot Lake were as different as the dark side of the moon.

Where 300 men worked underground at Dome, 1,000 worked in a mine at Elliot Lake. The Kid had never lived away from home; now he found himself in a huge bunkhouse full of strangers. At Dome everyone knew everyone else and was able to stop and pass the time of day. At Elliot Lake he was lost in the mad scramble, and no one knew who he was, or cared. The mine had to be producing a certain tonnage by a certain date; either a man could perform his assigned task, or he could not. No one was going to take time to show him the tricks of the trade.

Those big paychecks were not for any farmer that wandered in through the gate. If the stope boss told the Kid to drill and blast an eighteen-hole slash, it had better be done. It mattered nothing that he had to drag all the necessary gear from the bottom of the stope to the top. (The stopes were long rooms sloping up the flattish vein.) It did not matter if the rock overhead was loose. It did not matter if the other shift had scraped out the muck that he needed to stand on. All that mattered was getting that slash blasted. If the Kid could not do it, others could who were bigger, stronger, and more experienced than he. If the cross-shift started asking who was the bozo that drilled that slash, it did not make life any easier.

In the Kid's stope at the Dome, whoever was driving the scooptram slowed down when he saw a man's light ahead of him and stopped to see who it was and if he needed anything. At Elliot Lake the Kid had to flatten himself against the wall as the huge ST-8s rolled by. The drivers did not care who it was and reckoned, correctly, that the guy would climb the wall rather than be run over. The air was black with diesel smoke, and the Kid vomited his breakfast most mornings.

At Dome the Kid's shiftboss had known his father forever. When he came through the stope, he would say, "Hi, kiddo, how're ya makin' out?" followed by an appraisal of the Kid's work and a word on how to do it with greater safety or less effort. At Elliot Lake the shiftboss had an ever-changing crew of

thirty-five men, and the Kid was just "that noo #$%ˆ& they
drug in offa the #$%ˆ&* street. @#$% if I know where they get
'em." When he came through the stope, he just shone his light
on the face where the Kid was drilling, shook his head, grunted,
and left.

Life in the bunkhouse was not much better. The Kid bought
a motorbike and a stereo on credit. Then he got tangled up with
the gambling crowd, older and wiser than he, who expertly
cleaned him out. Back at Dome, the Kid's father had a word
with the mine superintendent and told the Kid that he would
bail him out, provided he came back to the mine. The Kid reckoned it was a pretty fair deal all around.

Besides being large and gray, the Dome resembled an ele-
phant in another respect: it had a long memory. Joe was another
miner who reckoned he would give it a go at Elliot Lake. Before
leaving the Dome, he put in a few hard shifts and drilled his
entire stope ready for blasting. The shot holes were supposed
to be eight feet deep. Not until after he had quit and had been
paid bonus for all the holes he had drilled did the company
discover that they were only four feet deep.

Like many others, Joe found that Elliot Lake was not all it was
cracked up to be, and he eventually returned to Dome. As afore-
said, the Dome had a long memory; the overpaid bonus came
off Joe's first bonus check after his return. He was known as
Shorthole Joe ever after.

Few Dome miners went to work at other mines in the Porcupine, and the opposite was also true, although as mines like the
Hollinger, Coniaurum, Buffalo Ankerite, and Paymaster closed
down, their former employees gravitated to the surviving mines
such as Dome. Dome miners never felt the need for a union
until after the Hollinger closed in 1968. Large numbers of Hollinger men, thrown out of work, hired on at Dome. The Hollinger had been unionized, and it was not long before Dome went
the same way.

Under the union agreement Dome could hire and fire whom-
ever they pleased without reference to the union, provided that
all new hires agreed for the company to deduct union dues
from their paychecks. Conversely, the union could not prevent
a man from being hired by refusing him membership. The contract contained a grandfather clause, however, whereby the
union could not collect dues from men who had been Dome
employees before the mine was unionized unless they chose to

join. In the 1970s a remnant of preunion miners remained who hated the union and anything to do with it. Threats from union hotheads and hints of possible intimidation did nothing to change their attitude, although little or nothing came of such threats.

As the years went by, the number of employees who had worked at the mine before it became unionized dwindled by natural attrition. With every contract negotiation, the union pressed for the abolition of this grandfather clause, but the company stood by its preunion employees, come hell or high water. In the end the number of men still exempt from paying union dues was so small that the company yielded, but only to the extent of obliging these men to pay the equivalent of their dues to a church or recognized charity of their choice. One or two diehards refused to be dictated to in this matter by anyone for any reason and quit or retired.

Texasgulf offered higher wages than any of the gold mines and, in its early days in the 1960s, had sucked miners from them like a sponge. By the middle 1970s, Texasgulf was regarded as a bunch of smart alecs, and the internal politics were reputed to be ferocious. Many of the men who went there came back to Dome muttering about "too many bosses" and other vague discontents. Texasgulf was strongly antiunion and successfully resisted all attempts at union organization. The union—United Steelworkers of America—pointed out, with some justification, that the threat of unionization forced Texasgulf to pay a better package of wages and benefits than the unionized mines and that the nonunion miners, whether they liked it or not, were freeloaders, owing much to the union but giving nothing in return.

In the summer of 1979, Dome experienced its first strike since 1913. After slumping in 1976, the price of gold rose in the next few years to undreamed-of heights. The Dome was making a healthy profit and unilaterally, in the middle of a two-year union contract, raised hourly wage rates by a dollar an hour, which represented an increase in the order of 15–20 percent. At the same time, they issued every employee one share for every two years of service. For a man with thirty or forty years of service, that was no mean gift. Even so, at the end of the contract, many of the men wanted more than the company had given them or was prepared to offer and went on strike.

No one really knows if a strike is going to happen until it

actually does so and the picket lines go up. At six o'clock on that bright, hot summer morning, John was standing on the sidewalk with his lunchpail beside his feet, hands in his pockets, waiting for his ride out to the mine.

One of the miners was known as Bibitt. If Bibitt was shown a ton of explosives in fifty-pound sacks and told to pack it up a hundred-foot raise, sack by sack, that is what he would do, without question or complaint, until the job was done. Bibitt was that kind of man. Bibitt had a big, shiny, black pickup with chrome wheels. As John stood on the uneven concrete slabs of the sidewalk, he looked along the street in the direction of the mine and saw Bibitt driving back into town very slowly. As he drove by, Bibitt raised his hand in greeting. That was how and when John knew that the strike was on.

The strike had the air of a holiday. Most of the miners scattered into the bush to go fishing and enjoy the sun. The mine staff put up with plenty of good-natured ribbing as they went through the picket line to patrol the deserted workings and check on the safety of the mine; sometimes cans of beer were offered to them. After two weeks boredom began to set in. For the first time, a paycheck failed to appear. That was the psychological moment when the company pulled a rabbit out of the hat with some small concession; the contract was settled, and everyone went back to work.

Each mine in the Porcupine camp was its own clannish community. Dome people were only vaguely aware of their counterparts at Pamour or McIntyre and regarded their opposite numbers as the dull-witted crew of a cheapskate employer. Occasionally Dome or Noranda employees would meet each other in bars or at mine rescue competitions and would realize, with a faint sensation of surprise, that "the other lot" were people just like them. John worked in the Porcupine for four years. In all that time, he never went underground in any other full-sized mine in the camp.

The Dome was not unique—special, yes, unique, no. All the way across a thousand miles of wilderness, from Pickle Crow to Larder Lake, other similar microcosms played their part in the gigantic mining industry of northern Ontario.

14

THE MIGHTY DOME

The focus of our attention is the Dome mine—greatest of them all, the Big D, the Mighty Dome—also known, not uncommonly, as "This @#$%&* Place." The mine and company were so called because of a dome of white quartz, splashed with gold, which had been the first discovery back in 1909. For most of the mine's history, the main shaft from the surface had been no. 3, marked by its headframe painted a certain shade of dark red known locally as Dome red. No. 3 shaft had been collared in 1915 and now ran vertically from the surface to about 2,400 feet, with levels every 100 or 150 feet. Around the shaft station on the 2,000-foot level was a network of tunnels not unlike a subway station, electrically lit and echoing with the shuffling of rubber-booted feet, shouts, the metallic clangor of muck trains, and the pervasive hiss of leaking compressed air. A half-mile ride in a steel-box mancar, drawn by a trolley-line locomotive through a dimly lit tunnel led to another subway station, also hewn out of the gray-black rock and whitewashed, surrounding the top of another shaft that began and ended underground.

This shaft, no. 6, measured 19′ × 13′ in rock—similar dimensions to no. 3. It began on the 2,000-foot level and went down to 4,000 feet. All the paraphernalia of skip dumps and sheave wheel supports, built of steel in the headframe at the top of no. 3 shaft, were here carved out of the rock, as was the cavernous hoist room. No. 6 shaft had been sunk between 1936 and 1939 to give access to deeper ore. Another slightly smaller shaft, no. 7, went down from the 4,000-foot level to nearly 5,300 feet from the surface, but at the time of this narrative it was disused and flooded.

The levels driven from the two main shafts, 25 of them, ram-

bled for miles. Few indeed were the men who could claim to know them all and to recall the whereabouts of each numbered drift, raise, and crosscut. Many, if not most, mines have some sort of layout that is repeated from level to level in response to the distribution of the ore. At Dome, however, the geology was such that the drifts on each level spread out like a vine on a trellis—25 levels, 25 vines on 25 trellises.

The deepening of the mine, which had gone on at intervals since 1910, had ceased with the sinking of no. 7 shaft in the 1960s. The methods and equipment available up to then had skimmed off the best ore, over 30 million tons of it. Improved mining methods and machinery and higher gold prices made it worthwhile to go back and mine the large tonnages of lower-grade ore that still surrounded the original stopes. In the late 1970s, therefore, work was in progress all over the mine, from the bitter cold 7 level, a mere 700 feet from the surface, to the pleasantly warm 29 level at 3,900 feet.

These labors were producing a steady 2,800 tons of ore each day. Each level was gradually extended in search of more ore, and the 9' × 9' drifts branched out ever farther into the rock like the probing roots of a great tree. Where ore was found, 7' × 5' raises were driven upward from one level to another to explore it in greater detail.

At this time about twenty-five stopes were at work, each one lasting for years or even decades, although some were worked intermittently according to the price of gold and the availability of men and machinery. Ore mined in each stope was dumped down a millhole, which led to a control chute on the level below. Ore was drawn through the chute into mine cars, holding four tons each, and hauled out to the shaft in trains of three by battery locomotives. There the ore would be dumped down the ore pass to where a crusher lay in wait for it in a chamber cut out near the bottom of the shaft.

The crusher would crunch it into chunks small enough to run easily through the skip loading system and discharged into another chamber known as the skip pocket. At the bottom of the skip pocket was a set of chutes where the skip tender dumped loads of rock into the skips as the hoistman lowered them to him one after another. Each shaft contained two skips that partly counterbalanced each other. The crushing and hoisting of the day's production began at 3:00 P.M. and went on until the early hours of the following morning.

At the bottom of each shaft the skip tenders worked in a small room beneath the skip pocket, electrically lit and much encumbered with steelwork, pipes, and valves. It contained the skip chute controls, a telephone, a wooden bench polished by much use and upholstered with well-worn burlap, and a grubby logbook with a stub of pencil. In this confined space the skip tender spent eight solitary hours each shift. The crusherman's facilities were similar.

The work of the crushermen, skip tenders, and rock hoistmen was comfortable enough, but dull, solitary, and almost entirely nocturnal. They would be coming underground as the day-shift miners were going to the surface. One miner might ask another, "Who's that guy?" as a skip tender headed for his rathole under the skip pocket, but for the most part they went about their business unnoticed and unsupervised.

A muck train hauled by a trolley locomotive shuttled between the two shafts and moved ore from the top of no. 6 shaft to the bottom of no. 3. When all the ore passes, skip pockets, and bins in the mine were full, they held a total of 12,000 tons, or six days' feed for the mill. This surge capacity was prudently managed so that it never became completely full or completely empty.

The company's geologists would identify an ore body by means of diamond drills that probed like blind worms hundreds of feet into the solid rock. Their wormcasts consisted of inch-thick cylindrical "core," which the diamond drillers pulled from each hole as they drilled it. This core was packed into boxes, labeled with the reference number of the hole from which the core came and the footage down the hole, taken to the surface, and analyzed for rock type and gold content. Miles of core were pulled every year.

To the casual observer it was all blackish rock, shot through in places with veinlets of white quartz. Here and there the veinlets glistened with patches of sulfides. Once in a while there might be a tiny speck of an unmistakable deep, rich yellow that seemed to glow in its stony matrix. That was gold. Most of the gold was in particles too small to see, so "VG" in the core log meant two things—"Visible Gold" and "Very Good."

The company geologists saw rock that was grayer or greener or bluer than its surroundings. They saw contacts and flow horizons and rock types that, in their strange jargon, were more mafic or more felsic. They found by assaying that some parts of

the rock mass carried that yellow impurity in slightly less in-
finitesimal amounts than others.

The geology and workings of the mine were plotted on plans
and cross-sections on which the geologists drew dotted lines
around where they thought the ore to be. On one side of the
line was rock that supposedly was carrying more than fifteen-
hundredths of a troy ounce of gold to the short ton of rock and
that would probably be worth mining. On the other side of the
line, it probably would not pay. Inside the ore body were tons
of rock without a scrap of gold in them, while outside it were
gold-bearing veinlets that would be left behind, either for some
future generation of miners or forever, because it was not worthwhile to mine through the barren rock surrounding them. As
the price of gold changed, so did the definition of what was ore
and what was not. These ore bodies were amorphous masses of
rock of a million tons or so, distinguishable only to the prac-
ticed eye from the waste rock surrounding them. In addition,
the mine contained ore in the form of continuous and clearly
defined gold-bearing quartz veins. Mining these veins cost more
per ton than mining the large ore bodies, but their high gold
content made it worth the extra cost. The company's diamond-
drilling program was five to seven years ahead of production.

The choice of mining method depended on the size and
shape of the ore body. Two predominant methods were in use:
cut-and-fill and blasthole.

Where an ore body was to be mined by cut-and-fill, raises
were driven through it from main level to main level, and an
exploratory drift, known as a sill drift, was driven at an elevation twenty-five to thirty feet higher than the main level at the
bottom of the future stope. Starting at the sill drift, the stope
was "silled out" to the width of the ore so that mining could
begin. Main-level drifts, sill drifts, and raises were driven by
specialized crews who moved about the mine with their equipment as required. After a stope was silled out, a stope crew and
the necessary machinery would be moved in to mine the ore.
Each stope was numbered according to the number given to the
drift beneath it. Thus 1372 stope was developed from 1372 drift
on 13 level, which was about 1,550 feet from the surface. Drifts
were numbered in accordance with an obscure logic.

In the process of mining by cut-and-fill, slabs of ore ten feet
thick were blasted down from the roof over the whole area of
the stope, which corresponded to the outline of the ore at that

elevation. The mining of each ten-foot "lift" of ore might take a year and produce 50,000 tons of rock, from which it may be deduced that some of the cut-and-fill stopes were quite extensive. This mining was done by sections, called panels, rather than throughout the whole stope in one piece. The blasting of a panel would leave a mountainous pile of broken rock. Two men would scramble up on top of this to bar loose rock from the fresh roof and secure it with rockbolts.

The panel would be mucked out with a diesel scooptram, or with a similar device powered by compressed air known as a Cavo, which dumped the ore down a millhole. A timber fence would be built across the panel, and the floor would be raised ten feet toward the roof by pouring in a watery slurry of sand, which was the waste product of the mill, conveyed underground through a series of pipes and boreholes. This material would drain and compact to form a new working floor; a small amount of cement was added to the fill for the top foot of each pour to provide a firm floor for machinery to run on. The millholes were built up with timber cribbing, leaving an open cribbed structure running up through the fill for the passage of ore.

The efficient management of a stope of this nature devolved on the mine captain in whose territory it lay. It was a complicated process requiring the shrewd orchestration of men, rock, and machinery. Some stopes worked well; others were the bane of anyone who had anything to do with them.

Some of these stopes ran continuously for fifteen years or more, expanding and shrinking in horizontal layout and extent in accordance with the shape of the ore body. Commonly half a dozen men would work in a stope on each of the two production shifts. Unless a man quit or was reassigned elsewhere, he might work in the same stope with the same fellow miners for years on end. The stopes had lives of their own, like some weird soap opera played out in the darkness, accompanied by the smells of blasted rock, explosives, and diesel smoke, the hammering of rockdrills, and the dull booming of blasts.

Then there were the blasthole stopes. Once an ore body had been defined and deemed suitable for this method of mining, the company spent years driving access raises and slot raises, drill subs and undercuts, scram drifts and drawpoints. For several more years, the cold, foggy drill drifts reverberated to the shattering roar of the longhole drills punching hundreds of two-inch blasting holes up to sixty feet fanwise into the ore. After years of labor and the expenditure of hundreds of thousands of dollars, they began production blasting. Explosives by the ton and rock by the thousand tons went west in cataclysmic blasts. The broken ore spilled out into the drawpoints; the scram drifts shook to the growling and bellowing of the yellow-painted, dirt-blackened scooptrams.

A scram consisted typically of a 10′ × 14′ scram drift from which half a dozen 10′ × 10′ drawpoints tapped the bottom of the stope fifty feet apart. A scram might be on a main level or between levels. In either event the scooptrams dumped ore into a millhole that had been driven as a raise from the next main level below. At the bottom of the millhole was a chute where trammers drew ore into muck trains for haulage to the ore pass. As fast as the scooptrams dug out the ore that spilled down into the drawpoints, more spilled down to take its place; at least that was the theory. Blasted rock was generally well broken up, with few pieces bigger than a cooking stove. But rock caved off the walls and roofs of the caverns, and beyond a certain point, they grew of their own volition.

Caved rock could vary from huge to enormous. If a drawpoint measured 10′ × 10′, then a twenty-foot slab was not going to come out too easily, and no one could handle it if it did. The standard treatment was a two-pound slab of explosive, known as Sakpak, laid on top of the rock. Sometimes a fifty-pound case or more was considered appropriate, and the results took out ventilation doors hundreds of feet distant, collapsed metal oil cans, and covered everything with dust. It would have been more economical and less destructive to drill the rock and blow it up with a few sticks of dynamite. However, there was no way of knowing if a big rock contained unfired explosives from a longhole blast, and it made no sense to find out the hard way.

The bigger the rocks, the more likely they were to hang up in the drawpoints. It could be a hazardous business to clamber up into a drawpoint to blast a hang-up because it might move and roll rocks down onto the unfortunate miner. If the rocks were very large, voids twenty or thirty feet high might develop inside the drawpoint. If a bad hang-up developed, the idea was to leave it for a week while mucking from some other drawpoint. If it had not moved in that time, it was probably safe to climb up inside and place explosive charges to blast it down— probably.

One time, John passed through 924 scram to find the crew staring at a single huge rock that completely blocked a drawpoint. The next drawpoint but one was empty. Warily, he scrambled up into it. The powerful beam of his hat lamp revealed nothing but darkness, so huge was the open cavern above. Looking across to the blocked drawpoint, he could see what the problem was. A single rock pinnacle, ten feet thick and forty or fifty feet long, stood upright with its foot in the drawpoint.

Sometimes they came larger than that. In 1328 scram all six drawpoints hung up. The cause was found to be a piece of real estate 300 feet long and 100 feet thick. Up above the scram a drift opened out into the big hole, through which the scram crew could clamber out onto the top of this monster. In considerable danger from further rock falls, they stacked explosives on top of it, a thousand pounds at a time, and blasted it until it fell apart.

Anything that would fit into a muck car was fine. The cars were reasonably capacious—but so were the rocks. One summer night, when half the crew were on vacation, a shiftboss needed two trammers. He had two summer students working for him. He asked if either knew how to blast; one said that he did. Toward lunchtime, a miner came out to the shaft and found the two students at the ore pass with a rock jammed in the car. They were barring and prying at it to no effect. "Why don't you blast it?" asked the miner and went to eat his lunch in the lighted shack by the shaft. Soon afterward, an immense bang blew the lights out. After giving vent to the exclamations that are customary in such situations, the miner poked his head out of the shack. The beam of his hat lamp showed the shaft station to be full of smoke and dust, which was billowing in clouds from the drift leading to the ore pass. In the smoky darkness, he nearly fell over the two students, who were sitting against the wall, white and shaking. As soon as he could penetrate the choking fumes, he groped his way to the ore pass. The rock had gone; so had the car door. The remainder of the car was a twisted wreck, resembling one of those metal sculptures that proliferate on the grass outside municipal buildings. Returning to the two students, he asked out of curiosity how much powder they had used. "One of those bags" was the answer. "One of those bags" was a fifty-six-pound sack of ammonium nitrate blasting agent.

To keep the rock spilling out of the drawpoints, two crews of three men spent their time humping sacks and cases of explosives and boxes of detonators, cleaning out the blasting holes when the time came to blast them, loading them with explosives, priming them so as to fire in the correct sequence, and firing the longhole blasts. As ore was blasted and drawn, so the excavations grew in size. The longhole blasters worked around these black holes. Sometimes the task was not conducive to any great tranquility of mind.

Some of the time, the longhole blasters would be working around one of the "big holes" while the scram crew were mucking the drawpoints below them. As long as the rockpile in the cavern covered the drawpoints, the scram crews could safely blast any oversized rocks without danger to the longhole blasters above. If one or more drawpoints was empty, the scram crews were told not to blast until the very end of the shift, and the longhole blasters made themselves scarce in good time.

One longhole crew was working on a plank staging several hundred feet above some empty drawpoints. The event happened in slow motion. The longhole crew saw a ball of fire blossom out of one of the drawpoints. As one man, they threw themselves off the staging onto the ground only a split second before the blast ripped their hard hats from their heads, slapped their faces, boxed their ears, and punched their chests all at once. The earsplitting crack of the detonation was followed by its long, reverberating roar. They groveled on the floor with their hands over their heads, trying to protect themselves from the debris that rained about them. They picked themselves up and looked at each other, shaking with fright. Much of the staging had gone; some of the remaining planks hung by single nails. They all knew what had happened; the idiots in the scram had piled their oversized rocks in the empty drawpoint and blasted them. Maybe that was why Mike Croke—who was on that crew—donated his Crescent wrench to John one day and went into the hotel business in Toronto.

During his first winter in the Porcupine, John worked for a short time (too long a time, in his estimation) on a longhole crew with Bill Miller and Jojo. The three of them got along well as a crew, but John disliked being a powder monkey—besides which, the longhole crews worked continuous day shift, thus offering no relief from getting up early in the morning.

When longhole blasting went well, which it did most of the time, it went very well, like the time when the three of them spent a week cleaning holes and another loading them with seven tons of explosives. That Friday they fired the blast; a rumble signified the conversion of 15,000 tons of rock into dust and rubble, and that was all. When longhole blasting did not go well, it could be a nightmare.

There were jackpots and death traps—and then there was 1517 slot. The layout that awaited them was really quite simple, like the drawings supplied to them by the earnest but dull-witted gnomes of the engineering office. They climbed fifty feet down a ladder through a raise into the slot crosscut, which was a chamber in the rock eight feet high and twenty feet wide. The ladderway came down through its roof into the blind end of the chamber. The other end, fifty feet away, opened out into the "big hole"—the cavern that had already been blasted out. Floor and ceiling were perforated with row upon row of blasting holes drilled upward and downward from the slot crosscut. It should have taken a week to clean and load the holes. On Friday they should have fired the blast—Pow!—and headed for the Legion—easy.

They thought they had better check the roof for loose rock before going to work. John tapped it with a scaling bar; the rock was drummy. (Sound rock would ring.) The experienced ear can tell by the sound, not only whether the rock is loose or solid, but how big and how loose a loose piece of rock may be. Wherever he tapped the rock, it gave out this same drummy response. Starting under rock that sounded more or less solid, John found a crack, jabbed the point of the bar into it, and pried. The rock creaked but nothing happened. He took a fresh bite with the bar and pried again. A small rock fell out of the roof ten feet away and smacked into the dirt. If a man barred in one spot, rock was not supposed to fall from the roof ten feet away. John pried once more. Like a bolt of lightning, a slab fifteen feet long, six feet wide, and two feet thick crashed to the ground. John sprang back, tripped, and landed on his back on the rough, rocky floor. He sat up, cursing and rubbing himself, while Bill picked up the bar and took his turn.

They spent days barring down loose rock until it became obvious that they were getting nowhere. By then they had barred four to six feet of loose rock off the roof, and this material lay in piles all over the floor, burying the blasting holes, which had to be loaded. There was so much rubble on the floor that the company decided to move a slusher hoist down into the slot crosscut, but even so, it took a week to get the place cleaned up. Then they had to build a timber staging so as to be able to reach the blasting holes in the roof and load and prime them. The outer end of the staging had to be cantilevered out into the cavern. Their safety lanyards became entangled with everything, but the alternative was to risk a one-way trip down into the darkness.

By and by, the holes in the roof were all loaded, and the primer tails hung down in their dozens. John thought that the simplest procedure would be to leave the staging in place. The blast would reduce it to matchwood. But someone in authority said that it had to be dismantled.

As they removed the staging, John was sitting astride a timber, his feet hanging down over the black void. Close above him was the roof, cracked through and through, and stuffed to repletion with high explosives and live detonators. John happened to brush it with the end of a plank as he passed it to Jojo. A one-foot cube of rock fell, rolled off his shoulder, bounced off his thigh, and sailed off into the darkness. Many seconds later, they heard the echoing crash of its fall. John eyed the roof thoughtfully. If the whole thing was as loose as that, they might be lucky to see daylight again. They removed the rest of the staging without mishap.

After they had loaded and primed the blasting holes in the floor, one task remained. Some of the rock forming the floor at the open end of the slot crosscut had slid off into the big hole, leaving an irregular cliff face overlooking the abyss. Some blasting holes were left ten or fifteen feet down this cliff face and still had to be cleaned and loaded. If they were left unloaded, the whole blast might fail. Someone had to go down the cliff, push a stiff polyethylene hose down each hole so that the mud and water left from drilling could be blown out with a blast of compressed air, load explosives and primers into each hole, and connect the primer tails to the trunk line of detonating cord. Seven or eight holes had to be dealt with in this manner.

John looked around. Bill was lame. Jojo looked as if he wished he were somewhere else. That left John. "Hell, I'll go," he said, with a bravado he did not feel. They hung a short ladder down the cliff, roped to an eyebolt in the wall. John roped himself to another eyebolt. Bill took the slack in the rope as John clambered down the cliff face.

Jojo lowered the end of the plastic hose to him. By waving his hat lamp, John could signal to Jojo to turn the compressed air on or off. As his freedom of movement was restricted, John could do no more than shield his face from the geyser of mud and water that the compressed air blew from each hole and that gushed all over him. Sometimes evilly disposed persons would piss in the holes after they were drilled, for the especial discomfiture of the longhole crews. John was spared the effects of this particular vile practical joke. When the holes were all clean, Jojo pulled the hose back up and went to fetch the explosives.

John looked about him. The cliff was in front of him; the floor of the slot crosscut was about ten feet above the level of his head. The walls of the slot, where it had already been blasted out, were about ten feet away on either side. Behind and below him was darkness, where he could hear the muckpile quietly grinding together as the scram crew drew it down into the drawpoints. Directly above him hung that awful mass of riven rock from which the primer tails hung down by the dozen. Behind and above him was the scene that terrified him. As it was blasted out, the slot had undermined the collapsed wreckage of an old shrinkage stope from forty years ago. A mass of rocks and timber was locked together fifty or sixty feet above him. It might be locked together so tightly that the crack of doom would not move it; but then, too, it might not. That echoing crack, far off in the darkness, might be just a small rock falling off the wall, or it might presage a collapse that would sweep him into eternity. He had no way of knowing.

Bill's light shone down at him. "You alright down there?"

"Yeah, just admiring the scenery."

A chuckle, then, urgently, "Come on, Jojo, get the lead outa ya @#$%ˆ& ass! Ya shoulda had that crap ready before! We wanna get the @#$% outa here!"

At last, down came a sack of powder cartridges and detonators on a rope. Jojo handed John the loading stick. To load each hole John slit a cartridge open, dropped it down the hole, and heard it plop on the bottom. He followed it with three more sticks and tamped them with the loading stick to burst the cartridges and compress the explosive into a solid mass. Several more sticks followed, also tamped, to load the hole to within two feet of the collar. He reached into the sack for a detonator, dug a nail out of a shirt pocket, punched a hole in a cartridge, stuffed the detonator into the cartridge, and tied off the detonator tail around the cartridge. He lowered the primer down the hole, followed by two more cartridges for good measure, and gently but firmly tamped the whole thing. Last, he looped the excess length of detonator tail together and tied it off.

The job could not be hurried or fumbled. Exposed as John's position was, he could not help wondering if each second might not be his last. He worked on, sweating slightly. Finally, he connected the primer tails to the main line of detonating cord, and the job was finished. He opened his mouth to get Bill to take in the slack in the rope while he climbed back up the ladder, but only a croak came out. He swallowed and tried again with better success and climbed back up to the slot crosscut.

They hauled their gear up the ladders to the level and sat down to wait for the end of the shift. An hour later, they fired the blast, and a rumble signified the conversion of their handiwork into fumes, dust, and rubble.

For a time, John worked with a graduate of the Haileybury School of Mines, a farm lad from southern Ontario known to his friends as Muffin. They worked under the tuition of Boris, one of the mine's training instructors. The work they were given to do consisted of driving a sill drift, plus various other odds and ends around the mine. The sill drift reverberated like an organ pipe to the thunder of the jackleg, which shook them to the bone. John had run various models of jackleg at one time or another, but this particular make seemed determined to break his back or his heart or both.

Given a choice, a jackleg driller will choose the most powerful machine he can lay hands on; more power means faster drilling in harder rock and hence more bonus. But increased power means increased weight, which is limited only by the average strength of the users for which the machine is designed. In North America both the jacklegs and the miners that use them are among the heaviest and most powerful in the world. Some lighter models of jackleg weigh 80–100 pounds, including the pusher leg, but these monsters weighed 130 pounds, evidently designed for some miner built like a tank slugging it out with the iron-hard rock of the Canadian Shield. Not without reason; the Sudbury nickel mines in those days would not hire a man for underground work if he weighed less than 150 pounds. But whoever designed these machines never had to use them, for they were awkward and ill balanced, which further increased their apparent weight.

Some safety whiz had decreed that the machines had to have a muffler that would resist all attempts to remove it, not caring if a man damaged his back because of the weight of the machine if a few decibels would be prevented from reaching his ears. The effect of the muffler, if any, was not apparent to the user, who wore earmuffs to protect his ears anyway. As well as adding to the weight of the machine, the metal muffler would ice up if the compressed air supply contained any moisture and would prevent the machine from running at all. It was by no means unknown to find a man almost weeping with frustration, screaming the vilest abuse he could think of at an iced-up machine that made only a futile hissing noise, while pounding it with a sledgehammer with all his strength. Of course, the manufacturer's salesmen smiled whenever an order for spare parts came in. In one stope six machines were needed to keep one running. After inflicting these wretched machines on their miners for nearly a decade, the company wised up and went over to a different manufacturer.

In any case, John had to do the best he could with the tools he was given. Boris was 205 pounds of solid bone and muscle and handled the machine like a toy. Muffin pushed weights in his spare time and was in the habit of doing squats with 400 pounds across his shoulders, leaving John, at six feet in height and 175 pounds, the smallest and weakest of the three. As John was wont to remark, the only miners who do not sweat are the very strong and the very idle; he was neither. So John and Muffin drilled and blasted, mucked and rockbolted, sweated and swore, and thus learned their profession.

Boris was a conscientious teacher and reckoned that the mine sanitation system was a necessary part of the curriculum. The standard installation was a box of waxed cardboard, with a supply of lime and sawdust, and a two-by-four for a seat. One day, while they were eating lunch, Boris described how, on one occasion, the two-by-four had broken beneath his weight, depositing him in the box, which had been well used at the time. He was quite offended when his two pupils rolled about on the floor, helpless with mirth.

The boxes were taken to the surface by a man who appeared on the time sheets as "scavenger" but who was commonly known by a more obvious appellation. The mine employed two scavengers, one working from the no. 3 shaft, the other from the no. 6 shaft. Both were Italian and hated each other so passionately that they could not be allowed within sight of each other.

The scavenger's position was a curious one, unwanted and for that reason privileged. That task was his whole and sole occupation, and he could never be roped in to assist in any other work. His job entailed less responsibility and less physical labor than almost any other. He was his own master and during the shift was free to come and go about the mine as he pleased. He spent much of his time wandering about alone or just sitting still and gossiping with anyone who might be so inclined. He was the eye that saw without being seen, the ear that listened without being heard. He was everywhere and nowhere. Get on any cage, walk any of the main drifts of the mine, and there you would find him, the wandering shitman.

The government mine safety authorities had decreed that the simple and effective device noted above was not good enough, and the company was obliged to construct monstrous contraptions of stainless steel and concrete. Naturally Boris had to show one to his pupils, commenting with a mixture of amazement and disgust, "I think they spent thirty thousand bucks on this @#$%^& thing." The Thing was indeed marvelous to behold. Boris threw open the steel door with a flourish. There, to everyone's embarrassment, was one of the timbermen enthroned in splendor.

These employments, with their humorous diversions, were far from random in either nature or intent. Almost all the shift-bosses and mine captains had worked at the mine continuously since returning from military service in the late 1940s. A good proportion had worked at Dome or at one of the other mines in the Porcupine before the war. In the late 1960s the company's senior management foresaw a time, still ten years in the future, when all their mine supervisory staff would retire. They foresaw, too, that mining would become a more complex business and that its management would demand more in the way of formal technical training. The days were gone when a man who learned to read and write in the bunkhouse could rise to be mine superintendent through sheer competence and force of character.

In mining, for generations, if not for centuries, there has been a division, often an acrimonious one, between the practical man and the technical man—if you like, between the empiricist and the theoretician. The miner with technical training or the mining engineer with dirt under his fingernails represents an ideal much sought after but seldom found. The Mighty Dome set itself the awesome task of producing that ideal by hiring technical school graduates, or graduate mining engineers if it could find them, making miners out of them, and then promoting them to fill the ranks of the retiring shiftbosses and mine captains. Muffin and John were but two of the trainees going through this system. Others, corresponding to each year's crop of technical school and mining college graduates, were making their way in the world of the mine with varying degrees of success.

The proof of the pudding was how well a man made out as a shiftboss. The dividing line between success and failure was extraordinarily simple. Those trainees whose physical strength and attitude to their work was such that they became competent jackleg drillers never had any trouble as shiftbosses. Those who were weak, lazy, or thought that running a jackleg was beneath their dignity and intelligence, never had anything but trouble in supervising the work of others.

The performance of these new supervisors, men in their middle twenties, was compared directly with that of men with thirty or forty years of underground experience, with ten or twenty years as supervisors. Those graduates of the program who left the Dome came to realize how demanding was the school from which they had graduated and what an incomparable apprenticeship they had served. From the company's point of view, it was sadly ironic that the most successful graduates discovered, almost without exception, that their attributes were in demand in more challenging and more rewarding situations than anything offered by the Mighty Dome.

15

A RUNG ON THE LADDER

The company promoted John to be a shiftboss. About 150 to 200 men worked underground on each of the two production shifts—7:00 A.M. to 3:00 P.M., and 7:00 P.M. to 3:00 A.M. The four-hour gap between shifts allowed the ventilation system to clear the mine of smoke and dust from blasts fired at the end of each shift.

Each shift was supervised by six or seven shiftbosses; each shiftboss was assigned a section of the mine known as a beat, extending over several levels, of such a size that he could visit every working place in the first half of the shift. The second half of the shift was spent filling in time sheets and other documentation, arranging for maintenance work, revisiting parts of the beat, checking on sump water levels, operating the ore pass control gates to run ore down through the system, and doing anything else that needed doing. At the end of the shift each shiftboss left a written summary in a logbook for the benefit of his opposite number, who had the same beat on the other shift. The cross-shift was under no obligation to follow suggestions from his predecessor or do anything that he asked him to do; inevitably, some shiftbosses got along better with their counterparts than others.

Only a man with a death wish would ignore the comments and instructions written in the logbook by his mine captain. Expressed in terms that were always blunt, and sometimes obscene, these notes from time to time attained a perfection of terse and unequivocal style that inspired awe in all present, especially the shiftboss in whose logbook they appeared.

Some beats contained two or more large cut-and-fill stopes, each with a crew of half a dozen men; the shiftbosses assigned to these beats had crews of up to twenty-five men. Other beats, al-

though of similar or greater territorial extent, were less active; shiftbosses assigned to them might spend the whole first half of the shift walking their beat, seeing only six or seven men in that time.

The production shiftbosses were responsible for stopemen, scram crews, driftmen, silldriftmen, raisemen, trammers, and a few laborers and mucking-machine operators. On the day shift, in addition to the production crews, the mine was populated by timbermen, pipefitters, riggers, trackmen, longhole drillers, longhole blasters, diamond drillers, electricians, mechanics, samplers, surveyors, and geologists, who reported to their own supervisors, as well as the more senior staff of the mine. All of these wandered at will through the shiftboss's beat. The shiftboss had no control over them yet was responsible in a general sort of way for their safety. At best, they got in the way of his production crews. At worst, he had to divert men and equipment to serve their needs, and they plundered his beat like a horde of locusts.

The new shiftboss learned his job by walking the beats with the regular shiftbosses and learning their routine. After two weeks or so of this training, he was considered ready to stand in as a replacement shiftboss. The maximum demand for replacement shiftbosses was in summer, when most of the regular shiftbosses took their vacations. After standing in as a replacement shiftboss for a time and otherwise doing whatever work was assigned to him, the new shiftboss would take over a beat whose shiftboss retired, was promoted, or was assigned to other supervisory or technical duties.

Life as a new shiftboss was far from easy. A man with ten or more years of mining experience was at an advantage over someone in the position of John and his contemporaries whose experience had been brief by comparison. Yet, when all was said and done, it was a long step from working in a stope, drift, or raise to organizing and coordinating a dozen or two dozen men and several different working places. Every new shiftboss, without exception, flapped and floundered, forgot things, and made ghastly and embarrassing mistakes. It was part of the price of taking that first step up the ladder.

The qualities needed to make a successful shiftboss were not easily definable, nor was it easy to predict correctly whether they would blossom or fade when a man made the transition, whether he would grow in stature or crumple under the burden of responsibility. Skill as a miner was no indicator, nor was intelligence, popularity, integrity or the lack of it, or even the gung-ho attitude of a sports team captain. Firmness, energy, and resourcefulness were part of the picture, but not the whole picture. Some men had what it took; some did not. By the same token, not every competent shiftboss had the necessary qualities to make the next step to mine captain, although the reason for that distinction was yet more obscure.

As soon as a shiftboss walked in through the door at the start of his shift, problems flew at him from all directions, to be averted or solved in a short space of time, in coordination with neighboring shiftbosses, using the men and equipment thought to be available at the time and keeping work priorities firmly in mind. This was no time or place for shrinking violets, prima donnas, polite convention, or formality. The clock ticked on, crews reported for work, seldom complete, and the cage took them underground whether the shiftboss had his affairs in order or not.

Every shiftboss had to make snap decisions on the basis of imperfect information leavened by sheer intuition; he had to make these decisions in the brief time available and accept responsibility for them. A bad mistake could endanger men's lives. The successful shiftboss made more right decisions than wrong ones—it was that simple. In later years some of John's colleagues and superiors were irritated by this habit of making snap decisions. What really irritated them was the realization that the decisions that they reached in twenty minutes were not substantially more correct than those that John reached in twenty seconds. Less popular still was John's offensive attitude that people who showed any form of indecision were losers, incapable of cutting the mustard on production, and fit only for engineering, where they could play with their pencils all day long without getting in any one's way. These traits were the imprint of the school of hard knocks that was production supervision at the Dome mine.

Organized or not, the crews disappeared underground and scattered to their various working places or sometimes wandered about in search of equipment. The shiftbosses went down after the last cageload of miners. Any miner who happened to be late for work had to take this cage under the baleful glares of half a dozen shiftbosses, which in itself carried the suggestion that he had better mend his ways.

Each shiftboss got off the cage on the top level of his beat. He spent the first half of the shift walking the levels from working place to working place and climbing down ladders from level to level. In each stope, scram, drift heading, or raise he would chat briefly with the crew and check on the safety of the working place and its equipment. He would confirm that the crew knew what they were supposed to be doing and had all the equipment and supplies they needed and would check that their machinery was functioning properly. Depending on the nature of the working place, he would spend from ten minutes to half an hour there before moving along on his travels. Most crews were friendly, a few were not, but that did not matter, because the foreman's job is no popularity contest.

If the shiftboss was in each working place only once a shift, it can be seen that his supervision of the work was highly indirect. What got the work done was incentive bonus and the work ethic. If a man was completely useless, he was not going to be around for very long, but such people were rare.

Development headings earned the highest bonus, at $25–$30 a shift, with most stopes in the $15–$25 range. Plenty of stopes and scrams, however, paid only $5–$10 a shift in bonus. If a man's base wage was $50–$60 a shift, he was not going to bust a gut for an extra $5, so that left the work ethic, the hope of better bonus next month, or the hope of assignment to a job that paid a real bonus. Such an assignment was obviously not going to fall from the sky onto the idle or the incompetent. If a man hired on as a "mucker" (as laborers were known) at $5 an hour, he had the chance, over a period of years, to advance to be a driller at $7.50 an hour and $25 a shift bonus. He could thus double his wages in constant dollars by making a good miner out of himself.

In the Porcupine in the late 1970s, $85 a shift was a comfortable living. It was known that a man could make twice that by working for a contractor, but at the cost of working longer hours in worse conditions, generally in camps in remote areas, and with little continuity of employment. For a married man, that was no way to live. John's experience of camp life had been the bunkhouse at Yellowknife, and he did not reckon it was any way to live either.

And so the Mighty Dome churned out its 2,800 tons per day. Viewed from a distant point, such as the company head office in Toronto, everything went along fine. Although the head office types would hardly be the first to acknowledge the fact, this continuity of production was due in large measure to the endeavors of the shiftbosses.

Viewed from closer at hand, the task of supervising the work seldom went smoothly. Equipment broke down; minor emergencies occurred; 10–15 percent of the crew was always absent for one reason or another; upper management changed their plans unpredictably; events did not turn out as intended—all of which hindered the main purpose of the shiftboss's existence, which was to maximize the production of ore. If enough circumstances combined against him, a tough week on the day shift could seem like a surrealist nightmare in 4D and Eastmancolor directed by Murphy. Not surprisingly the following week on the night shift was eagerly awaited. On the night shift only the production crews were at work, and senior management was absent.

The position of shiftboss was the hardest and least rewarding of any in the mine, although some men stuck at it for decades. The shiftboss was between the hammer and the anvil, all too often hounded and abused by upper management, while being the prime recipient of every complaint from the miners. Everyone blamed everything on the shiftboss, who sometimes found himself being reprimanded for something over which he had no control before he was even aware that it had happened.

The classic attributes of the successful shiftboss included an abundance of low cunning, broad shoulders (figuratively and preferably physically as well), a thick skin, a ripe sense of humor, and a foul, sharp tongue, for this was a rough-and-tumble business where a man was expected to stand on his own two feet. Whatever threats management might breathe, however, the mine would not function for many hours without the shiftbosses. Throughout John's time no man once promoted to that position was ever demoted or fired.

Many of the miners made more money in wages and incentive bonus than the shiftbosses were paid in salary. A miner could not be ordered to work overtime, but if he chose to do so when overtime work was available, he was paid at one and a half times his normal hourly rate. A shiftboss could be ordered to work overtime, yet was paid only at his regular rate for doing so. The shiftbosses had to be at the mine well before the crews arrived because of the horse trading that had to go on between them to set things up for the shift. They did not leave until the last man was checked out from underground and the last scrap of paperwork was complete. Their salaries did not take these ten-hour shifts into account. On the day shift, that meant getting up at 4:45 A.M. Under those conditions John could never sleep properly at night; he struggled through the week on two or three hours of broken sleep each night, dragging himself to work each morning feeling as cold and sick as if he had a hangover. The local radio station put in its two cents' worth at that time of day with two songs—"Take This Job and Shove It" and another that contained the line "She doesn't have to get up in the morning," which left him muttering and cursing in disgust.

It seemed that everything that could go wrong did so on the Friday day shift. Sometimes things would start to come unglued on Thursday. If the process began as early as Wednesday, the prognosis for Friday was grim indeed. One Friday at lunchtime, after a tough week, John was filling in his crew's time sheet in the room that served as a shiftboss office on the surface. He managed to get the thing started but kept losing control of the pencil, which scrawled off, self-willed, into parts of the time sheet where it had no business being. Otherwise he stared blankly at the printed lines and columns as if he had never seen them before, his mind knocked out by lack of sleep. He woke up ten or fifteen minutes later to hear the assistant mine superintendent telling him, "We have to work overtime tomorrow, so I guess you'll be working day shift."

Each shiftboss worked for a mine captain. Three mine captains were in charge of production, and one was in charge of timbering and construction work. John worked for Eddie Richardson. Rock-hard and leather-tough, a more honest man than Eddie Richardson never stepped in shoe leather, nor was there ever a more competent mine captain. Eddie expressed himself with a degree of obscene blasphemy that John never heard equaled by any other human being. He kicked John into shape in short order. The kicking process was more informative than enjoyable, but it taught John a number of lessons that he never forgot and that could not be learned in any other way. Tony Horbul, the shiftboss on the neighboring beat, also worked for Eddie Richardson. He briefed John succinctly: "Whatever Eddie says, do it. If he tells you to go around the corner and stand on your head, you go around the corner and you stand on your head." (Except that the metaphor used was considerably less polite.) "Don't talk to him or draw sketches on pieces of paper; he hates people that do that." (The hatred of Eddie Richardson and the wrath of Smaug had points in common.) "All he wants out of you is yes or no—and it had better be yes."

The shiftboss's job was, essentially, to convert the mine captain's wishes into reality, to ensure that the conversion took place in a safe and efficient manner, and to report that this had been done. As time went on, John learned that certain set-piece situations repeated themselves and had corresponding set-piece solutions. Beyond that, a shiftboss was expected to keep things running smoothly and ensure that minor problems did not intrude on the mine captain's time or attention, as the mine captain was responsible for the activities of 80–100 men on two shifts, compared with the shiftboss's 10–20 on one shift. If the mine captain had to pick up the pieces for one of his shiftbosses too often, that shiftboss was apt to get his horoscope read.

The job required little in the way of originality; the decisions to be made were limited in scope and repetitive. John found that it was no use planning for or worrying about the morrow. When the shiftboss walked into the mine office at the start of the shift and read the reports left by his cross-shift, those were the cards dealt to him to play as best he could.

The relationship between miners and shiftbosses in those days was a friendly one. Nevertheless, a gap of sorts inevitably existed between them. When John crossed it, he was leaving the secure comradeship and simple responsibilities that he had enjoyed as a miner to prove himself in new and often trying circumstances. As time went on, he found a closer-knit fraternity among the shiftbosses that had its own brand of raucous humor.

The new shiftboss was invariably subjected to the most unmerciful teasing at a time when he was still unsure of himself, hog-tied by problems of a kind he had never faced before and was not sure how to solve. Most people learned to laugh it off and gave as good as they got. No holds were barred, and sometimes the original humorists found themselves outclassed in what was repaid to them.

The company had plans for John's career that, from the company's point of view, were grandiose. They involved rapid promotion to secure and well-paid positions of responsibility that would carry prestige in the community. Perhaps, even, as a man approached retirement after serving as mine superintendent or mine manager for twenty years, he might be picked for that coveted promotion "down to Toronto." This was, after all, the right and proper career for a mining engineer. From the narrow perspective of the company and the community, this idea was true and valid. The snag was that John did not have that same perspective; besides which, he was at heart a tramp miner.

In the depths of one winter John was watching television with his friends Cec and Lorraine Hobin. All was snug inside the house, but outside, the snow on the lawn was up to the windowsills. The starry night was a crisp thirty-five degrees below zero, and the temperature would probably be forty below by morning, or colder. The show on television was a football game televised live from California, played outdoors in the soft, golden light of a warm evening. Most of the spectators wore shorts and T-shirts. Dressed like that in South Porcupine, you would be frostbitten before you could sprint across town, and you would be dead and frozen stiff before daybreak. To live like this was absurd.

All about him John heard the voices of failure. Even the voices of success were the muted tones of those who had settled for what they had. Indeed, the very act of leaving South Porcupine was in itself regarded as a token of success. Tales came back of friends and relations who had gone to Toronto or the States or "out West" and had done well for themselves; those who had not were the ones who returned. There was a better life to be had, and John meant to have it. His resolve hardened with the passing seasons.

This was easier said than done. The mining industry all over the vast expanses of the Canadian north offered the same monotonous landscape, the same interminable winters or worse, and the same way of life as the Porcupine, but for the most part in places even less attractive. At 5:30 on a winter morning the differences between Yellowknife, Thompson, Manitouwadge, Timmins, Sudbury, Kirkland Lake, or Rouyn-Noranda are so slight as to be imperceptible. For that matter, we could include Norilsk, Vorkuta, or Karaganda, and the differences would not be startling.

Far away beyond the Prairies was a never-never land called British Columbia. In its mountainous interior there were mines, but from the viewpoint of northern Ontario, they were either vast open-pit mines or small, haywire underground mines, here today and gone tomorrow. The fact was that, from northern Ontario, it was a very long way to anything different.

The "good jobs"—John knew that they existed without being quite sure what or where they were—were snapped up by people who were in the right place at the right time and whose experience was longer and more varied than his. No one was going to come looking for him, and he could pass all his days unknown in the wilderness of northern Ontario. There was, indeed, an insidious temptation to let fatigue and, as time went by, monotony take their effect, to arise monklike before dawn to make his oblation at the mine and, taking the path of least resistance, to let time slide by. John was to find, as did others, that to break away from northern Ontario took a ruthless, unquenchable resolve and a will of iron.

16

THE ALARM-CLOCK KID

The position of mine captain was the best in the whole hierarchy of the mine. The incumbent was responsible for the work and safety of two shift crews totaling eighty to one hundred men, large amounts of machinery, and the safe and efficient extraction of substantial tonnages of ore. His jurisdiction covered roughly one-third of the physical extent of the mine. Within reason, the mine captain had a free hand to take whatever decisions and make whatever arrangements were necessary. At Dome in those days, the mine captains ran the show.

The mine captain was the most senior official directly concerned with the nuts and bolts of the work, yet good shiftbosses would take care of most of the day-to-day details and minor irritants. Conversely, a good mine captain knew when and how to take the load for his shiftbosses. At the same time, the mine captain was insulated from the time-wasting and obnoxious outsiders with whom the mine superintendent had to deal. The mine captain's rank was one of great and ancient authority and accordingly was much respected.

The company thought well of John as a shiftboss and promoted him to be the night-shift mine captain, a position that he relished because, like the heroine of the song, he did not have to get up in the morning. Unfortunately, the company saw to it that he did not escape the day shift entirely, but at least when he was on days, the promotion gave him an extra forty-five minutes of sleep in the mornings.

On the night shift John was responsible for everything that went on underground during the night hours. He bore this responsibility with an easy mind because of a thoroughly compe-

tent crew of shiftbosses with whom he spent many an agreeable shift.

The situations that confronted the night captain were extraordinarily varied. Some were humorous; some were hair-raising.

One evening at the start of the night shift, John was sitting by the telephone that connected the shiftboss office with the outside world while the shiftbosses stood at their wickets checking their crews and assigning work. The incoming shift milled about in the dry outside the office.

The telephone rang. One of the miners was on the line, saying that both he and his partner had colds and would not be coming to work. Perhaps there was a little too much noise in the background to be consistent with recovery from an illness severe enough to prevent the sufferers from coming to work. John, who suffered from colds incessantly and came to work anyway, received this information with no sympathy and scant courtesy. He told the man that, if he did not come to work, the fact would be recorded as an absence without leave and pointed out that his AWOL record was causing the company to lose interest in retaining his services. With that final snap, he hung up.

He retailed this news to the shiftboss concerned. Tony Horbul, one of the more cynical members of that cynical crew, suddenly burst into such mirth that he was speechless for several minutes and tears ran down his face. The words that came out in gasps and squeaks between fits of laughter ran, "You stupid sonofabitch! The guy phoned from the bar. They're both stinko, and if they ever get here, you'll have to send them home again!" The joke was on John that time, and the shiftbosses headed for the shaft, leaving John to wait for the two miscreants, stropping his fangs and wearing a circular track in the concrete floor of the dry.

The two men finally appeared, dressed to go underground and gung-ho for a night's hard work. Their glassy eyes and an almost visible aura of beer fumes showed that this was not to be. Both were leaning in different directions. Whereas before, they had not wanted to come to work, now both they and John were equally annoyed when he told them to turn around and head back to town. The two men decided to make a night of it and headed for Timmins along the back road, where they rolled their truck into the ditch.

The problem of men coming to work drunk was a persistent one, mostly on the night shift. Only the hardest of hard cases came to work drunk in the morning, but a hangover passed unnoticed in the grungy ranks of sick, sleepy, foul-tempered individuals that assembled out of the predawn darkness of a northern Ontario winter morning. It is also a fact that some men work better when slightly drunk than they do when completely sober. The symptoms of drunkenness could easily pass unnoticed, unless they were extreme. But the mine had a way of looking after its own and "Went home sick" looked better in a man's records than "Sent home drunk." What was dreaded was a drunken hoistman who could easily kill someone; regrettably, such instances are by no means unknown. Intoxication by drugs other than alcohol was occasionally suspected but never proved because no one knew anything about the symptoms.

"Lunchtime" on the night shift was at one o'clock in the morning. The shiftbosses, having traveled their separate beats, forgathered in their offices to eat lunch and fill out their time sheets. Besides the office on the surface, there were crudely furnished, electrically lit cubbyholes with telephones on the 2,000-foot level near the bottom of no. 3 shaft and the top of no. 6 shaft.

As night captain, John made it his business to cover as much of the mine as possible in the course of the week. This time, he was eating lunch in the office on the surface with the shiftbosses who looked after the upper workings. The shiftboss on the top levels at that time was a friend of John's by the name of Paul Wright.

"Y'know," said Paul, "those guys working in nine-twenty-four sub have got to be crazy."

"Why?" asked John.

"Right by the edge of the big hole, there's a crack all the way around the drift."

There were places in the mine—not many—where the rock was moving so much and on so large a scale that cracks appeared all around the walls, roof, and floor of a drift. If the rock in 924 blasthole stope was moving like that, it was serious.

The cogwheels went around in John's head. The longhole blasters were working in 924 sublevel, cleaning holes for a blast, working on the day shift only. The leader of that particular crew was an inveterate drunk, but he was by no means stupid. The clear implication was that the crack had appeared since the day shift.

John and Paul reached the sublevel around 2:30 in the morning. The rock that was being longhole blasted was itself the wall of a cavern that had been mined out decades before. The drill drift—924 sublevel—opened out into a black hole. Twenty feet back from the edge of the void, a gaping crack extended all around the walls, roof, and floor of the drift. Paul said that it had grown wider even in the past few hours. A colossal slab of rock—John reckoned later that it could have weighed 50,000 tons—was in the process of toppling into the void. Even now, it was making little noises. All of a sudden, very slightly, very deliberately, the whole thing moved. The two men fled. When something like that let go, there was no knowing what might follow. The slab fell sometime before the day shift. The longhole blasters were faced with the hair-raising task of clambering about over a pile of immense boulders, stuffing explosives into whatever remnants of blasting holes they could find, and blowing them up.

Any mine foreman, but a mine captain in particular, could suddenly find responsibility thrust upon him, immediate, heavy, and direct, even the responsibility for men's lives. One morning on the day shift John was accompanying Morty Mortensen on his beat. They were passing through 2415 stope when someone came hurrying up to them with the words, "Larry Jones went down the millhole." Larry Jones was the stope boss in 2506 stope, where the crew were building up the timber cribbing of one of the millholes in preparation to receive backfill. Normally this was done when the millhole was full of muck, on which the crew stood as they worked. In this case the timber boss had asked for the millhole to be drawn empty so that the timbermen could repair the chute at the bottom. The millhole had been pulled empty, but the timbermen had delayed repairing the chute, and now the millhole had to be built up to allow that part of the stope to be backfilled. The stope crew had built a plank staging inside the 8′ × 8′ millhole and had proceeded with the work.

John's immediate reaction to the information given by the breathless miner was, "Well, what did he do that for? And what of it?" A split second later, the truth dawned on him. Larry Jones had just fallen fifty feet down the empty millhole.

Ten or fifteen minutes later, they were in 2506 stope. John looked down the millhole and could dimly see a human form motionless at the bottom. A few questions revealed that the staging planks inside the millhole were not nailed down. Larry Jones had flipped one out of place accidentally and had fallen through the gap.

Some millholes had manway compartments built into the timber cribbing structure beside the millhole itself; this one did not. The cribbed section of the millhole was vertical for fifty feet. The bottom twenty-five feet below that ran at a fifty-degree inclination through the rock sill pillar beneath the stope. At the bottom was a chute built of timber and steel. The control gate was formed of seven or eight anchor chains hanging down from a steel beam in the brow of the chute, yoked together, with a row of eight-inch steel balls hanging from the yoke. This assembly was moved up and down by a compressed-air piston to release ore into the muck cars.

They scrambled down a manway in another part of the stope and walked along no. 25 level to the chute. Larry had apparently landed on the sloping rock bottom of the inclined section of the millhole in an upright posture and now half lay, half sat, with his knees against the chute chains. The groans coming from his inert form showed that he was at least alive.

In such a situation, people materialize out of nowhere. Besides John and Morty Mortensen, there was Ernie Morden (the backfill man), Randy Delgiudice, and Walleyed Gauthier. John understood with discomfort the sudden sharp division between the leader and the led. As mine captain, he was the leader; the others clustered around him with an urgent, unspoken desire to be led, to be told what to do, and to set about it knowing that everything was under control and would work out all right. This time they had a badly injured man on their hands; they only knew that he was not dead—yet.

John turned to Morty. "Go back to the shaft; telephone first aid; tell him to get the quack and come down here with a stretcher and a first aid kit. Then you stay on the shaft station and guide them in here." No. 25 level was 3,350 feet from the surface, and the rescue party had to come down no. 3 shaft, over to no. 6 shaft, and down to the level, so help could by no means be expected immediately. Morty set off down the drift.

"Ernie, go get a short piece of ladder. Nab it from any place you can find one. Ray, go and find two pipe-wrenches so we can undo the chute chains." Ernie and Walleyed Gauthier disappeared on their errands.

Randy was thinner and lighter than John, so John squatted

down and locked his fingers together. Randy put one foot
into John's clasped hands, the other onto his shoulder, and
scrambled up onto the front of the chute. There was just enough
give in the heavy anchor chains for Randy to squeeze in between them. John dragged himself up onto the chute and did
the same.

"Where are you hurt, Larry?"

"All over," came the semiconscious reply.

"Which bit hurts, then?"

"Everything hurts."

Randy and John looked at each other and shrugged their
shoulders. John pinched Larry's shin. "Can you feel this?"

The fact that Larry had feeling in his leg meant that at least
his back was not broken. They checked him for bleeding and
broken bones but found none. They had between them a semiconscious man, in pain, but with no readily identifiable injuries. They put their jackets over him to keep him warm and
talked to him to try to keep him conscious. It takes about five
seconds to say, "You're going to be OK. We're going to get you
out of here soon," to a man who has obvious and profound
doubts as to the truth of either of those statements. You cannot
go on saying that kind of thing forever; at the same time, other
topics, such as the weather or last night's hockey score, seem
somewhat lame. So Randy and John spent the next half hour
repeating the general message of Larry's imminent removal to
more comfortable surroundings while he groaned with pain,
sliding in and out of consciousness and repeating in slurred
accents, "Get me out of here. Get me out of here."

Morty Mortensen, the first aid man, the doctor, Ernie, and
Walleyed Gauthier converged on the scene all at once, with a
locomotive, a stretcher, tools, and a first aid kit. They unbolted
two anchor chains from the yoke and spread them apart to allow the stretcher to be pushed inside the chute. With some difficulty, they maneuvered Larry into the stretcher, which caused
him great pain, although they were unable to identify any injury, internal or external. They slid the stretcher out onto the
top of the locomotive.

Zorba arrived. Zorba was a spare mine captain of vaguely
specified function. He had an extraordinary ability to place
himself in the limelight. On this occasion he had been in a
remote part of the upper levels. By means unknown he had
sensed where the action was and spirited himself to the scene

just as the stretcher was placed on top of the locomotive. Naturally Zorba took charge at that point and escorted the casualty to the surface, leaving John and Randy inside the chute to extricate themselves as best they could.

No one ever told John or Randy whether their performance had been good, bad, or indifferent. After the usual witch-hunt as to whose fault it had been, the episode was forgotten. Larry Jones must have been built like a tank. His only injuries were a concussion, bruises all over his body, and a cracked cheekbone. Mercifully, his memory was a blank between the instant of falling through the staging and coming to in hospital. Larry Jones was exceptionally lucky, for the mine is an unforgiving place for the unwary or the unwise.

Louie was a thoroughly good kid, active, intelligent, hardworking, and good-natured, as well as being the son of one of the company's most expert miners. Consequently, he had not been at work in the mine for long before he was running a jackleg, earning drillers' pay, and making bonus in 1372 stope. That was where he and John crossed each other's path.

John came to work one evening to be told that a blast in 1372 had uncovered a quartz vein in one wall of the stope that was richly impregnated with high-grade. John's job for the night was to have someone drill short holes into the vein and blast them with a few sticks of dynamite. The high-grade was to be collected in sacks and taken to the surface for handing over to the security guards.

John got hold of Louie, who set up a jackleg and drilled a few holes, placed so as to slash the patch of high-grade off the vein, which was indeed pleasing to the eye. Even though they loaded the holes with half-sticks of dynamite and wooden spacers, the blast still scattered high-grade all over that part of the stope.

After an hour or so, they had collected most of it. Laden like pack mules, they clambered out of the stope and went to the surface. It was about the middle of the shift. By way of a reward for services rendered, John sent Louie home and made sure that he was paid for the balance of the shift, before sitting down to eat lunch with the shiftbosses in the mine office. The security guards took charge of the high-grade, which was thrown into the mill, where it contributed to the corporate bottom line.

A year or two after leaving Dome, when John was working in a different part of the continent, he heard that Louie had been killed underground. His death was hideously simple. He had been trying to bar down a brow of loose rock in a narrow stope. Because of the way the cracks ran through the rock, he was working underneath it. The shiftboss and mine captain happened to pass through the stope at that time; they told him in no uncertain terms to get out from there and showed him how to work on the loose rock from a safer position. No sooner had they gone on their way than Louie went back underneath the brow and must have started to pry at it once more. The whole mass of loose rock collapsed on top of him, killing him instantly.

The news preyed on John's mind; the gloomy horror of the mine bit him as never before. The rock was always watchful, silent, hard, and passive. It never said anything. But if a man made one false step, one unguarded move, it would reach out and kill him. There was nothing emotional about it, nothing about trapped miners, frenzied rescue attempts, or relatives standing about the pithead. Men went underground, and some were not quite wise enough or quick enough; in ones and twos the rock got them.

John was always on his guard; that was one reason why he was still alive. (Blind luck was the other, or divine providence, depending on one's point of view.) But he had often noticed that he was tired even if he came up from underground after no more than an hour. A man had to be fully aware, fully alert, as soon as he left daylight, otherwise he might not see daylight again.

The mine captains had amassed substantial amounts of paid annual vacation time by reason of their long service and exalted rank. Sadly, too, some of them were suffering the effects of too many hard years in the mine and were absent on sick leave for weeks at a time. The night captain had the additional function of standing in for the regular mine captains during such periods of absence. Sometimes shiftbosses or mine captains were unavoidably absent on short notice because of sickness or accident. Spare shiftbosses might be unavailable on short notice or at night, and in summer they were fully occupied replacing regular shiftbosses who were on vacation. The night captain, by virtue of his roving commission, knew the whole mine well enough to take over the territory of any supervisor—shiftboss or mine captain—on five minutes' notice. From the company's point of view, the night captain was like a tuxedo, not always fully occupied but essential when needed. If a man was appointed to that position, it was a harbinger of further promotion.

Agreeable though the position was, it did, however, entail closer contact with the mine superintendent, an individual whom John nicknamed the Iron Hand, in accordance with his style of management. The list of nicknames grew.

Stan Kukurin worked in the chill, dank caverns of 1016 stope. Stan celebrated his wife's birthday with a gift calculated to express his undying affection and esteem—namely, an alarm clock. The clock was large, cheap, and vulgar. When provoked, it jangled like a fire alarm. On close inspection, it proved to be of Russian manufacture. Stan's wife took an immediate dislike to the thing and told her bewildered spouse to get it out of the house. Stan took it to the mine and installed it, as much for decoration as anything else, in the corrugated-iron shack that served as a lunchroom on no. 10 level.

The workings extending from no. 3 shaft were for the most part cold and damp; only at 2,000 feet and below was the mine decently warm. In winter, ice was sometimes seen as deep as 1,000 feet from the surface. In one abandoned drift on no. 3 level, about 250 feet from the surface, a weird garden of ice stalagmites would grow from the floor every winter, some of them five feet tall and two inches in diameter.

Sometimes, in those upper levels, bats fluttered silently through the beam of a miner's lamp. The bats uncovered embarrassing phobias. Men who apparently feared neither God nor the devil would try to claw their way through solid rock to escape from someone who claimed—usually falsely—that he held a bat concealed in his cupped hands.

In the upper workings, the operators of the diesel scooptrams could hang cans of soup against the exhaust manifolds and eat their lunch seated on the engine covers, but otherwise the cold offered no encouragement to sit around. The lunchrooms by the shaft were lit with large electric bulbs that gave off a slight amount of heat. If half a dozen people crowded into some of the smaller shacks, they could become almost perceptibly warm—unless one of the Italian trackmen who ate garlic sausage happened to be working on the level, in which case the lunchroom did not stay crowded for long. The lunchroom on no. 10 level was large and cold, a bare, uninviting sort of place.

The Iron Hand saw the alarm clock in the lunchroom. His imagination conjured up visions of miners sprawled about, asleep on the company's time, relying on the alarm clock to wake them in time to catch the cage to the surface—a curious and no doubt unintended comment on the effectiveness of the company's incentive bonus scheme.

The mine captains were called together at the end of the shift. The Iron Hand closed the outer door of the office and, with the alarm clock as Exhibit A, expounded scathingly on their dereliction of duty, lack of interest in their work, so on, et cetera, and so forth. The shiftbosses in their office outside sniggered, as was customary on such occasions, glad that the hot water was being evenly distributed, as they considered their share inordinately generous. True to form, the door opened and the mine captains stalked out in close order, looking neither to the right nor to the left, ready to strike to the ground anyone unwise enough to inquire as to their feelings.

The alarm clock remained on display as an Awful Reproach for several days until John decided that he might as well add it to the collection, deployed around his apartment, which dragged him from sleep in the mornings, and packed it out in a brown paper bag.

The misfortunes that befell John over the next few weeks need not concern us here, except that they were numerous and varied. Not least, the young lady who was the light of his life at that time transferred her affections to another—someone who did not have to get up early in the morning and was therefore better company in the evening, someone too, it might be added, who did not have nasty, crude friends.

The light bulb went on in John's head at 5:30 one dark morning. He sat at his kitchen table, sick with the flu, trying to force down a meager breakfast, holding his head in his hands to prevent it from falling apart each time he coughed. He noticed the alarm clock watching him. Like Mrs. Kukurin, he suddenly got the message and resolved to harbor the thing no more under his roof. A drop kick into a snowbank would have solved the problem, but any object possessing so potent a jinx obviously had further valuable services to perform, so he took it to the mine.

That morning, John and Captain Shillington rode out to the mine with Eddie Richardson. Eddie's car squeaked and rattled in the fierce cold, lurching and crashing over the packed snow on the road out to the mine. Snowflakes passing through the headlight beams streamed at them out of the darkness. A red fuzz was the tail light of a pickup ahead. The snowbanks and the dark bush swept by half-seen. Not a word passed between them. The only human sound was that of Eddie Richardson spitting into a receptacle that the dawn's early light would reveal as a corroded tin can revoltingly encrusted with dried tobacco juice.

John set the alarm clock to ring at 3:00 P.M. and hid it in Captain Shillington's desk. The good-natured and guileless Captain Shillington was the butt of much of the humor in the mine captains' office. Three o'clock in the afternoon was the time when most of the crises erupted that demanded immediate attention by the mine captains at the exact time when everyone was legging it away from the mine as fast as they could go.

The mine captains' office was well provided with telephones, each representing some element of the system that had grown up over the years. There was even an army field telephone, with its own circuit to parts of the mine, which had been installed twenty years before and on which no one had thought to improve.

As the hour approached, John shooed everyone out of the office and stood idly in the doorway, watching Captain Shillington toiling at the paperwork on his desk. The infernal device cut loose with a jangling that would have had Mrs. Toschak's clientele throwing their tombstones at each other. Captain Shillington jumped straight up, overturned his chair, banged his knee, swore, and tore about the office, picking up each telephone receiver, bellowing, "Hello? Hello?" into each in turn and calling repeatedly on his Savior, while the racket continued unabated. In the end, he tracked it to its source and drew the alarm clock from its hiding place. Seeing John convulsed with laughter, he thrust the clock upon him with the spluttered command, "Will you for @#$% sake throw this @#$%ˆ& thing down the @#$%ˆ& ore pass before it hurts someone!" John sensed a widening awareness of the alarm clock's baleful influence and hid it in the mine superintendent's filing cabinet. The mine superintendent gained a new nickname—the Alarm-Clock Kid.

The company planned to sink a completely new vertical shaft from the surface to 5,400 feet. This not only would bypass no. 3 and no. 6 shafts but would enable the company to mine ore that they had tried to reach through no. 7 shaft back in the 1960s. At the same time they planned to increase production from 2,800 tons per day to 4,200. The mill was to be rebuilt, and the whole scheme would cost nearly $100 million. It would be the biggest single project ever undertaken at the mine and would rank as one of the biggest development jobs in Ontario for its day. The new shaft was to be sunk on a hummock of bedrock near the company housing, a short distance from the mine.

Speedball, Slingshot, and John could see no reason why the honor of cutting the first sod should go to some stuffed shirt from Toronto. Armed with a pick and a shovel one summer afternoon, they found the survey marks indicating where the new shaft was to be. They hacked a knee-deep pit in the stony soil matted with tree roots and photographed each other in appropriate poses. They threw the tools into the back of Speedball's pickup and headed for the Legion to make suitable libations and announce the good news that the first sod for no. 8 shaft had been well and truly cut. That evening there was even more than the usual amount of drilling and blasting in the Legion. The Royal Canadian Legion was ideally situated from John's point of view because it was within easy crawling distance of home.

The company had the last laugh. For some picayune reason dreamed up by the pencil-pushing nerds in the engineering office, they decided to move the shaft by a hundred feet or so, thus exemplifying the Golden Rule: he who has the gold makes the rules.

There was more to no. 8 shaft than just a hole in the ground. It symbolized the passing of the old Dome, with its family feeling and family feuds, as it had been for so many years. The old guard, the master craftsmen under whom John and others served their apprenticeships, retired.

The commissioning and early years of no. 8 shaft revealed technical glitches, some of which were of the most alarming description. By a stroke of ghastly irony Ed Sulis, one of the company's engineers, was presenting a paper on the no. 8 shaft hoisting system at a convention in Toronto when the news broke that a skip had fallen down the shaft. Fortunately, no one was hurt, but several months went by before the devastation was repaired. Meanwhile, maintenance had been allowed to lapse on the old shafts, and the mine was in dire straits from loss of production. Most of the underground crew had to be laid off until the repairs were complete.

The mine was plagued with union grievances, strikes, and fatal accidents in unprecedented numbers. As time went on, each strike lasted longer and bred more hard feelings than the one before. In 1987 Dome Mines Ltd. was merged with (some said taken over by) another company, and a new sign was hung up outside the gate. Matters did not improve. However, by the time these things ran their course, John was long gone in search of other adventures in a land beyond the sunset.

17

THE GOLDEN WEST

The year 1848 was an eventful one. In Europe it was a year of revolutions. In America, before January was out, an event occurred that was to alter the course of American history. John Marshall found gold at Adolf Sutter's mill.

In those days California was an extension to the Spanish jurisdiction of Mexico and had little to do with the English-speaking American nation on the far side of the continent except to supply it with rawhide. Richard Dana, an articulate Bostonian who went to sea in the 1830s for the good of his eyesight, spent many months on the California coast gathering and loading hides. His memoir, *Two Years before the Mast*, impresses us with the quietness of that desolate coast, baking in the heat of the sun. It sparked no fire in Dana, who longed desperately to return to Boston and get on with his life.

Adolf Sutter was Swiss and, so we are told, regarded himself as the benevolent despot of broad acres in the idyllic Sacramento Valley. When his foreman brought in some nuggets of a metal they recognized as gold, which he had found in the new millrace, the idyll of the unfortunate Sutter was destroyed forever. A horde of gold seekers swarmed over his lands, trampling and digging and plundering as they saw fit. They spread all over the valley and surged up the creeks to the east into the gentle foothills of the Sierra Nevada. At first they came from the port of San Francisco, but soon they were pouring in from "the States," both through San Francisco after a sea journey around Cape Horn or over the Panama isthmus and, after journeys of even greater hardship, overland from the east.

The swash and backwash of this flood spread through the mountains and desert of the American West over the next fifty

years, up through British Columbia and into the Yukon and Alaska. The first gold that the prospectors found lay loose in river gravels. Any fool with a shovel and a frying pan could stake a claim and mine it. Some people realized that the gold in the river gravels had to come from somewhere and traced it to veins in bedrock. At that stage the arcane skills of the hardrock miner came into play.

The hardrock miners, who originated in Wales, Cornwall, and Germany, were a different breed, apprenticed in a hard school and heirs to a corpus of knowledge going back for centuries. Plenty of the shovel-and-frying-pan brigade thought they could do it for themselves, and still do. They found it uncommonly hard work, or they lost their shirts, or both, and they still do.

The land that these people found was extraordinary. Almost everything about it was extreme. Its horizons were serrated by jagged mountain ranges that towered into the sky, whose peaks were places of bare rock and eternal snow. The clouds that swirled about them roared and spat with thunder and lightning. Winter blizzards deposited snow tens of feet in depth. Its deserts were loathsome spreads of rock and gravel, devoid of life except in grotesque and hostile forms. In places its native inhabitants killed without the slightest compunction anyone who trespassed on their lands. In this land a man could die of frostbite or sunstroke, thirst or drowning, avalanche, snakebite, or arrowshot with equal facility. It was vast, stretching from the mountain wall of the Rockies in the east for 1,000 miles to the Pacific coast in the west, and from Mexico for 2,000 miles and more into the Arctic. But above all, it was beautiful. It was incredibly beautiful, beyond the power of human telling. And whether they could express themselves or not, men loved it with a passion matched only by the savage strength of that violent land.

What lay beneath its surface was as incredible as that which was exposed to human sight. The ground was rich in minerals of all kinds—gold, silver, lead, zinc, copper, molybdenum, uranium, vanadium, tungsten, borax, coal, oil, and others besides. This was no mean mineral wealth, either. Like everything else about this land, its mineral wealth was extravagant, incredible.

This bounty gave rise to an immense mining industry that flourished especially during the last half of the nineteenth century. Wherever any sort of mine or group of mines came into being, a camp sprang up to house, supply, and entertain the miners. Most of the camps remained shack towns on bleak, windswept mountainsides and vanished as soon as the mines were abandoned. Some achieved a certain brick-built respectability and survive to the present day as mining towns, tourist attractions, or in some other function.

This land and this industry gave free rein to the inclinations of the footloose tramp miner. The two best jobs he ever had are the one he left and the one to which he is going; the worst job he ever had is the one he has now. The act of packing his gear and moving on is endlessly seductive and of potent symbolism. Perhaps nowhere in the history of mining has any part of the world given such ample satisfaction to this urge as the American West in the last half of the nineteenth century.

It need not surprise us that the golden West called to John in the wilderness of northern Ontario. No matter how warm and deep were the friendships that he found there, no matter how well people thought of him or how much the company valued his services, when all was said and done, underground production in northern Ontario was a hard, dull affair in a cold, bleak landscape. Time after time, as he looked out into the snowy, predawn darkness, he thought, "Life is not meant to be like this." Yet as long as he stayed where he was, it would never be anything else. Around him were people who accepted that life was supposed to be "like this," who openly disbelieved that anything better was obtainable. Some of them daydreamed fatuously of winning a lottery and retiring to Florida on the proceeds. John would never accept those ideas or surrender to them, nor would he ever agree with those who held them.

A chance meeting in Toronto one Christmas put John in contact with the American Borate Co., which was constructing a borax mine on the edge of Death Valley. Within weeks John was on board a Boeing 707 on the way to Las Vegas, Nevada, having taken a short vacation from the Dome to go and rustle a job.

The plains of Illinois and Iowa lay under deep snow. Across Nebraska the settlements became fewer, and the grid-line pattern of the section roads broke up. The snow became thinner, and in places the dun-colored land showed through. The fortress wall of the Rockies passed beneath the aircraft's wing. Narrow canyons wandered into the mountains, their snowy slopes furred with a dark stubble of trees. Tall mountains and upland parks gave way to the valleys and mesas of the western slope. Roads, railroads, and towns were few and far between, scarcer still where the aircraft cruised over the brick-red desert of Utah, dusted with snow and studded with rock towers and buttes. The plane dipped down over Nevada, and they landed at Las Vegas in the warm dusk.

The engineer at the mine had promised to meet John, who therefore squared his shoulders and marched briskly into the terminal building. No one was there to meet him. He wandered about like a lost dog, wilting under this abandonment. Owning a car was not necessary in the Porcupine, so John had no driver's license and now could not rent one. A convention was in progress, so no accommodation was available. John slept fitfully on an airport bench.

The next morning a young surveyor from the mine came to collect him, introducing himself as Wade. All troubles were soon forgotten as they bowled along a desert basin in the cool, bright morning. "We're just getting started, and things are a bit haywire right now," remarked Wade by way of explanation. John would learn to dread those words; when things started out haywire, so they tended to remain.

It was John's first experience of the American desert, or indeed of any desert, and he looked about him with an infinite curiosity. "View of distant mountains" would aptly describe this part of Nevada. The roads ran along the centers of intermontane basins five or ten miles wide, floored with brown and gray gravel and dotted with sagebrush. The nearer ranges could be seen as tumbled rockpiles baked into lifelessness by the sun's heat. The more distant ranges stood in mauve and blue silhouette against the paler blue sky.

"This is Lathrop Wells," remarked Wade as they turned at a fork in the road.

"Where?" asked John.

"This. Right here."

The words designated a water tower, beyond which a single trailer lay some distance back in the sagebrush. He went on, "You'd never think it, but there are ranches all along the edges of the valley. You can't see them by day, but you can see their lights any time you're along here at night." The feet of the mountains, with their spreading outwash fans, seemed indeed to be devoid of habitation.

"That's our mill," said Wade, pointing to a cluster of buildings under construction about half a mile from the road.

As they ran on for mile after mile across the desert, John's mind relaxed, freed briefly from the cold and darkness, the heavy responsibility of the mine, the nagging problems, the pinpricks of irritation and frustration that stung and bit day after day, month after month, and that would go on doing so for decades if he let them. For the moment, he was on vacation, soaking up what was going on around him with an alert but idle curiosity, an unfocused interest. The blue Nevada sky was in itself a promise of something better, whatever and wherever it might be.

They came to a low pass where the road ran through a cleft in the hills. "These are called the Funeral Mountains, by the way." Wade gestured at the black burnt rocks like slag dumps that now surrounded them. A half-built house perched improbably on a mountainside. "That was going to be the manager's house," Wade remarked, "but there was some kind of a screwup, and they spent a million bucks and only got it half built." John's antennae twitched. The road ran on through the jaws of the pass and down a long grade.

"That down there is the company trailer site." Wade indicated a patch of leveled ground some distance below the road. "The first families are already moving in," he added as though nesting boxes for some rare bird were attracting their first residents.

"How far is the mine from here?" asked John.

"About fifteen miles."

"Where does everybody live?"

It seemed incomprehensible to John that anyone would park themselves on a rock-strewn desert mountainside, remote not only from the mine but also from any settlement.

"Some of them commute from Las Vegas," answered Wade. "The rest live in Furnace Creek. That's about fifteen miles the other side of the mine from here." Las Vegas was 120 miles from the mine. "By the way," he went on, "you can't get anything in Furnace Creek. Everyone buys their groceries in Las Vegas."

Wade chuckled. "The local watering hole is the Furnace Creek Inn. One of the miners gets all gassed up and sees this guy leaning against the bar—doesn't know who he is. So he goes up to him and says he's going to punch him out. The guy says, 'You can't do that, I'm the mine superintendent.' The miner says, 'All the more reason,' and nails him one right in the kisser."

Furnace Creek was a motel and a scattering of trailers in a grove of tamarisk trees on one edge of the floor of Death Valley. On the far side of the valley the snow-capped Panamint Mountains towered 10,000 feet into the sky. Tamarisk trees were the only vegetation of any size that would grow in these conditions. They grew along a spring-fed creek that flowed out of the mountains and disappeared into the floor of Death Valley. Even the tamarisk trees were in danger because some government department claimed that they were not indigenous, having been brought in by earlier generations of borax miners, and that they should be removed.

They found the mine engineer still in bed in one of the trailers. Introductions having been made, John was taken to the motel and left to his own devices. After a bath, he slept the afternoon away. The mine engineer, Wade, and John gathered that evening for a convivial supper. Before going to bed, John wandered about Furnace Creek, filling his lungs with the mild night air. It was now February, and looking up at the stars overhead, he could not help thinking of those same stars shining down on the snowy wastes of northern Ontario, where even now it would be thirty or forty degrees below zero.

The next morning the mine engineer met John for breakfast and took him to the mine to show him around. The mine stood on the side of a draw. A shaft had been sunk for 700 feet, and a level had been driven out to tap the borax ore body that underlay the draw. It was not convenient to allow visitors underground that day. "We're just getting started, and things are a bit haywire right now." They visited a short tunnel driven into a hillside, in which lay another mass of borax. Instead of drilling and blasting, a machine was being used that chewed the soft rock out mechanically. The machine seemed to be in an indifferent state of health, and many individuals, including the manufacturer's representative, were in attendance at its bedside.

John admitted to the mine engineer that he had no intention of living in a place with so few amenities as Furnace Creek. He learned, however, that a vacancy would be forthcoming in the company's engineering office in Las Vegas and jumped at the opportunity. One of the mine draftsmen was driving back to Las Vegas that afternoon, and with mutual expressions of goodwill, John and the mine engineer went their separate ways. John returned to Canada the next day; American Borate paid his expenses in full.

In March the company offered John the job at Las Vegas, and there the matter stuck. For six months he rode a roller coaster of hope and despair as the company tried ineffectually to obtain immigration clearance. In the end he received a short letter from them to say that they had given the job to someone else. Whoever the someone else was, it did not do them much good because the mine closed down a year or two later. Rumors indicated that as things had begun, so they had remained.

John's struggle to escape from northern Ontario and move to the American West lasted for two and a half years. Finding work in the United States was a catch-22 situation. To obtain an entry visa, he needed a job to go to. No company would hold a job vacancy open for the time that it took to obtain a visa. The fact was that a junior mining engineer working as a mine foreman in northern Ontario had no special value outside that area. To obtain the more varied experience that would make him more attractive to employers elsewhere, he first had to escape. People like John were valued in northern Ontario for the precise reason that he wanted to leave; relatively few mining engineers were prepared to settle down to a twenty-year stretch in underground production in northern Canada. It was a tangle.

The fraternal grapevine of the mining industry broke this deadlock. One of John's friends, another Englishman, knew a Camborne School of Mines graduate by the name of Geoff Hunkin. Geoff had his own consulting business in Denver—Hunkin Engineers Inc.; he not only offered John a job but was prepared to wait for the immigration procedure to grind through its interminable process. Geoff offered John a job at the end of 1979 that he was able to accept a year later. In leaving the Dome, John left the corporate hierarchy and consigned himself to the wandering life of the tramp miner.

The snow had just started to fly when John left the Porcupine for the last time. As the aircraft climbed among the clouds one cold November afternoon, John caught a glimpse of the Dome mine—a constant feature of his life for the past four years. The clouds slid beneath them, and northern Ontario was gone.

John rented an apartment in a barracklike building on the southern outskirts of Denver. To the south the golden brown western plains stretched away to the hills behind Castle Rock, and on a clear day he could see Pike's Peak, a hundred miles away. To the north lay the outskirts of Denver. To the west the Rockies stood up like a wall into the crisp, bright Colorado air. To the east lay the flat, low horizon of the prairies. John no longer had to get up at five o'clock in the morning or rely on being on the night shift to catch up on lost sleep. The telephone no longer rang at three o'clock in the morning to ask for his instructions on some emergency or to ask him to come to the mine and solve some problem that the shiftbosses could not handle. John arose at a civilized hour and drove to work at the company's little office in Englewood.

Hunkin Engineers had branched out from mining engineering into the technology of leaching uranium from sandstone beds. From there the company had diversified into the obscure business of drawing groundwater samples from the innumerable boreholes that pierced the sedimentary strata of the western plains for various purposes. John learned a new vocabulary in which Smeal rigs and thorium shadows took over from the three words that had chiefly concerned him hitherto—drill, blast, and muck.

The company was developing new and improved groundwater-sampling equipment and testing it at well sites on the high plains of Wyoming. "Energy" was the big word in those days, and the state was rich in coal, oil, and uranium. Places like Cheyenne and Kemmerer and Rock Springs were busy day and night with the coming and going of drillers and truckers and geophysics operators. John and his new colleagues worked hard; sometimes it seemed that they worked all day and drove half the night as they traveled from job site to job site. Once, they holed up in Casper to rest and reorganize after a string of twenty-hour days. They ate steak for breakfast, steak for lunch, and steak for supper before moving on to the next job on the following day. All around them spread the rolling outlines of the high plains like a brown blanket spread over the earth's bones. In the distance, from time to time, they would see a range of purple mountains.

John went with Geoff Hunkin to supervise some exploration drilling on the sandhills near Andrews, Texas. Andrews is about fifty miles from the twin cities of Midland and Odessa on the west Texas oil patch. John learned that "the oil patch" was not a specific place but referred to wherever petroleum was produced, whether it was the Beaufort Sea or the Gulf coast. The oil patch was even less specific than that. Like the mine, it was a way of life, a state of mind, a challenge as fiercely demanding as the mine and often more abundantly rewarding.

Life on the oil patch went on at full blast 168 hours a week. John and Ed Oakes, a consulting geologist, often drove to or from the job site in the early hours of the morning. The sandhills were studded with lights, and on every back road they met rigs on the move or trucks loaded with pipe or sacks of drilling mud. The drilling was done by a contractor from Grants, New Mexico. The drillers worked twelve-hour shifts around the clock in blowing dust, the deafening blare of the diesel engine that drove the rig, and the clonking of the drill string. After their shift the drillers would head back to a motel in Andrews, where people like them were coming and going at all hours of the day and night. Longing for a cold beer, they could find none because Andrews County was "dry"—no alcohol.

Geoff Hunkin, Ed Oakes, and John lived in style in the Midland Hilton and drove the fifty miles out to the site every day. John and Ed found out all about the drinking habits of the inhabitants of Andrews County in the early hours of one morning. They shared a rented Pontiac, which had a fair turn of speed. On this early morning, for a change of scene, they joined the four-lane highway from Odessa to Andrews, with the Pontiac performing fully to specifications. Suddenly a stream of cars and pickups caught them up from behind, weaving all over the road, and overtook them as if they were standing still. They pulled off the road in alarm until this hazard abated. The good ol' boys from Andrews, being unable to imbibe in teetotal Andrews County, went across the line to Odessa. Ed and John had just seen them heading for home.

By day the low, flat horizon was interrupted chiefly by "nodding donkeys," the rocking beams of the sucker-rod pumps that drew oil from two horizons, 5,000 and 10,000 feet below the surface. There were big ones and little ones of all colors, shapes, and sizes, all nodding and wheezing and squeaking together. Sometimes they were in herds; sometimes there would be one all by itself, hidden behind a sandhill, slowly bucking and kicking its feet in the air. Some were dismantled and rusting, caught in midnod and disemboweled by wolves. In places the air was foul with the mephitic odor of mercaptan gases; but it was the smell of the oil patch, and hence the smell of prosperity.

One of the clients who contacted Hunkin Engineers was Vern Hughes of Twin Bridges, Montana. Vern owned a scattering of claims all over the Tobacco Root Mountains in the southwestern part of the state.

The Tobacco Root Mountains are a little-known area of past

gold mining. The peaks of the ranges rise to 7,000 and 8,000 feet; broad and pleasant ranching valleys run between them and around their ends. Twin Bridges lies in the middle of one such valley. The mines are hidden in canyons that wander back into the mountains.

In the 1860s the river gravels of Alder Gulch sprang into prominence for their gold content, and Virginia City, Montana (as distinct from Virginia City, Nevada), made its own contribution to the lore of the old West. It was a tough country in those days, remote from any centers of supply or transportation. Access was by trails whose courses through the mountains added many hard miles to their length. Deep snow and bitter cold threatened the very survival of those bold enough and tough enough to stay through the long, vicious Montana winters. June came before the sun banished the last snow from the mountains.

The prospectors and miners reasoned, rightly, that the gold in the river gravels had to come from somewhere. As some of them dug and panned and sluiced in valley bottoms, others began to find gold ore in bedrock under the trees. As the easier pickings of the river gravels became exhausted, the prospectors moved on and big operators moved in with dredges to rearrange the valley floors into arcuate ridges of dredge tailings. Hardrock mining demanded more money and fewer people than did placering. The boomers moved on, and the district became a hardrock camp. Towns like Pony, Norris, Twin Bridges, McAllister, Sheridan, and Silver Star grew up to serve the needs of men who worked in mines like the Boss Tweed, the Revenue, the Sunbeam, and the Elenora. In common with thousands of similar mines all over the American West, these faded away, leaving rock piles and adit mouths and weatherbeaten timber ruins.

The star of Vern Hughes's properties was the former Bielenberg and Higgins mine, a series of adits driven into gold-bearing veins at the head of a canyon in the mountains east of Twin Bridges. Vern's two sons took John in a jeep up a rocky road climbing along the precipitous side of the canyon. The road ended at a house where a young couple lived and kept an eye on the mine.

Behind the house was a timbered adit portal of generous dimensions. Extensive rock dumps indicated that the adit had penetrated far among the roots of the mountains. It had, however, caved in somewhere inside and did not allow access to the mine. In May the valleys and lower slopes were clear of snow, although the wetness of the gravelly soil showed that this was but recent. At 7,000 feet, however, snow still lay thickly between the trees and in other places where it had drifted or was shaded from the sun.

To go farther they had to put on snowshoes. Hoisting rucksacks onto their backs and hitching lamp belts around their hips, they set off up a snowy trail through the trees. The trail led past an old stamp battery whose remains stuck up through the snow. Kevin pointed out a square block like a gunsight, just visible far above them on the crest of a saddle. The block was a derelict cabin once inhabited by a miner who worked at the Bielenberg and Higgins. The cabin was fully 1,000 feet above the mine. The shift's work was enough for most men, but after each shift this miner had climbed back to his eyrie among the high rocks, where even now swirled rags of cloud that would become the afternoon's thunderstorms. When need arose, he would walk the five miles into Twin Bridges to buy supplies and backpack them to his solitary cabin.

The trees opened out around a snow-filled bowl in the mountainside. Kevin and Rick eyed the place and, picking a spot beneath a certain tree, burrowed into a snowdrift. They broke into the mouth of an adit leading into the mountain. The three of them crawled through the opening, and John found himself crouching on the surface of ice about four feet below the arched roof of a five-foot-wide tunnel. The adit had been driven slightly upgrade so that water would drain out. Snow and ice outside had dammed the water, so that John was now squatting on its frozen surface. The upward slope of the adit allowed them to walk gradually more upright the farther inside they went. The natural temperature of the rock was above freezing, and the ice grew thinner as they went farther into the mountain. The water was still knee-deep where the ice became too thin to bear their weight, and John gasped as icy water filled his boots. Plates of ice banged against his shins. After a few hundred feet they were on dry ground, but by no means dry-shod.

John had never seen such a rabbit warren of drifts and stopes and raises running in all directions. Alone, he would soon have become lost. He chipped rock samples from stope faces and drift headings and scuffed handfuls of muck from stope floors and old chutes. By the time they had toured the accessible parts of the mine, the rucksack on his back was well loaded. In many places they could go no farther because of collapses, or dared not go because of rotten ladders.

As John squelched along in his wet boots, the water in them became warm. They were, however, replenished with cold water on the way back out through the adit. In the process John had to climb onto a single slab of ice that occupied the whole width of the adit. The walls offered no handholds, and top-heavy with rock samples, he seemed assured of a somersault into the icy water as the slab tilted under his weight. By good fortune he crossed this obstacle without mishap and crawled out through the burrow in the snowbank to find the others squinting in the sunlight.

A journey of a mile or so on snowshoes remained to travel back to the caretaker's house. John with his burden lagged behind. A black bear hurtled across the trail between them, crashing through the underbrush. Much to their relief the animal showed no interest in them. John's feet were almost frozen by the time they were gathered in the cottage to dry their socks on the stove, drink scalding hot coffee, and swap yarns. Thunder rumbled in the mountains as they drove back to Twin Bridges. Vern Hughes treated them all to a feast in Dillon that evening, and with the best of mutual goodwill, John returned to Denver to have the samples assayed and to write his report with recommendations.

Unfortunately work of this nature was slow in coming, and the mining engineering work for which Geoff had hired John was elusive. In 1978–80 the price of gold reached extraordinary heights, especially when compared with the average cost of mine production. This spawned a wave of stock promotions. Promoting stock was easier and potentially more lucrative than mining gold. Many of the projects that mushroomed all over the American West in those days were "stock plays" pure and simple. Some others consisted of unholy combinations of fools being preyed upon by crooks, neither of whom was in the least anxious to be revealed for what they were by a competent mining engineer. Some schemes were ratholes driven into mountainsides by people who had neither the desire nor the money to engage a consultant to tell them what they thought they already knew. John launched a modest advertising campaign, but when Geoff saw the impecunious and disreputable characters who were attracted to his door, John's days were numbered. If there is one thing a consultant must have, it is a well-heeled clientele who pay their bills.

After only six months Geoff told John that his position would become that of an "associate" and that he was thus responsible for finding his own sustenance. John had foreseen this event, which merely placed the sharp edge of need on an employment interview that he had already arranged at a place only a few hours by road from Denver, a place called Cripple Creek.

18

A PLACE CALLED CRIPPLE CREEK

Bob Womack was an eccentric and solitary cowhand. In the 1880s he herded cattle in a high mountain valley around the headwaters of one of the creeks that flowed down into the Arkansas River. This creek came to be known as Cripple Creek, supposedly from the cattle that lamed themselves as they stumbled over its boulders. Indeed under a cover of coarse, tussocky grass the soil was little but a reddish gravel.

The country around Cripple Creek was open parkland 10,000 feet above sea level. The wind sighed in the scattered pines or hissed in the rough grass. Even a fit man would pause often to draw breath in the thin air. As he did so, he might hear the "Proink! Proink!" of a soaring raven printing its call on the quiet of the high meadows. In winter the bitter wind stalked like a swordsman over the hills. Snow fell but blew away or evaporated into the dry air. In summer the electric sky exploded with lightning. At night a man could read by the glare of a full moon. To the south the horizon was bounded by the Raton Pass, 130 miles away on the borders of New Mexico, to the west by the jagged peaks of the Sangre de Cristo Mountains at a distance of 50 miles, while to the north and east rose the brick-red shoulders of Pikes Peak.

There was magic in the hills, and not necessarily good magic. Nevertheless it was more than Bob Womack's gold that drew men to Cripple Creek, and that draws them yet. It was the gold that focused their energy onto the slopes of Globe Hill. Bob Womack knew the gold was there—and had known it for years—but not

until 1891 could he cut a sample that assayed rich enough to prove it.

Colorado had been gold-mining country, among the best, since 1858, and prospectors knew too well for their own good what Colorado gold looked like in the rock. The gold at Cripple Creek occurred as a telluride, silvery if at all visible to the naked eye, and in fractures in the rock less obvious than the white quartz veins that striped the hills elsewhere. That was why it went unrecognized for so long.

Cripple Creek, Victor, Mound City, Altman, Goldfield, and the other camps that sprang up all over the hills and gulches boomed with the mines. The whole district was only five miles across, but the pitting and tunneling and shaft sinking and drilling and blasting and timbering and hoisting that was concentrated in those few square miles in the thirty years after discovery was phenomenal. So, too, was the volume of water that flowed into the mines as their narrow shafts probed 2,000 feet and more into the granite and breccia.

Gold ore was rich and plentiful. Cities, replete with modern conveniences, grew up around the district, connected by roads and railroads. Some settlements were, and remained, raw, squalid shack towns straggling up narrow gulches. Cripple Creek and Victor, however, after suffering disastrous fires, grew up with foursquare brick buildings whose luxuriously appointed interiors housed the sophisticated life of a city. Railroads from Canon City and Colorado Springs brought the necessities of life and supplies for the mines, and also luxuries for which the generous harvest of the rock made such ample provision. Some parts of the district were photographed so often that it is possible to watch a camp grow up and then fade away.

The first photographs of Cripple Creek, taken soon after 1890, show the bare hills with a few tents and shacks dotted about at random. Next came functional but hideous clapboard timber buildings aligned on muddy streets. These gave way, in turn, to brick buildings forming the core of a city of 25,000 people spread over those same hills that had been devoid of habitation only a decade before. We move on to the present, and the last scene bears a curious resemblance to the first. The grid of streets still criss-crosses the shallow bowl in the hills, but now only scattered houses and clusters of buildings dot the upland meadows. The rest were abandoned one by one, were pulled down, burned, or fell down. Only a litter of stones and scraps of timber sticking up among the coarse grass and weeds tells where a house once stood. The buildings along the main street serve the tourists who come in from Iowa and Illinois and Indiana to gape at the remains.

Unknown among the tide of tramp miners that surged through Cripple Creek in its glory days was a young Cornishman named Eddie Tape. Eddie was not a miner by trade, nor did he ever work as one in Cornwall. In 1910 he went to Canada to work on a tobacco farm near Tillsonburg in southern Ontario. Then he joined the other Cousin Jacks in the copper mines of the Keweenaw Peninsular in northern Michigan. Next he lit out for the golden West and worked at the Portland mine at Cripple Creek before returning to England to fight in World War I. Surviving that, he became a carpenter in Cornwall and stayed there for the rest of his days.

Eddie Tape was an old man living with his daughter, son-in-law, and grandchildren in a snug house sheltered in the lee of Rosewarne Downs when John knew him. He kept with him some photographs, a piece of Michigan copper, and his memories. John listened, and Cripple Creek beckoned with a call he must one day answer.

One winter evening in South Porcupine, John was setting type at Bruce Watters's printing press. Bruce's basement was a good place to be when it was thirty below zero outside with four feet of snow on the ground. Bruce happened to show John his employer's annual report, which carried a brief mention of renewed work at Cripple Creek. The camp had been dormant for years, but there was fire in the ashes yet, now blown into life by the high price of gold. The sunlit hills of Colorado were infinitely more attractive than the boreal forests of northern Ontario. Once again the place beckoned. When his employment in Denver came to its abrupt end, John naturally headed up the hill to Cripple Creek. At that moment the company was looking for a mine foreman. The eye of fate winked, and the circle was closed.

The tide of mining in the district ebbed and flowed as erratically as the clouds that swirled about the mountaintops. There was still gold in the rock, and now there was gold in the pockets of gullible investors, and of companies, cash-rich after the boom years of the 1970s, looking for tax write-offs and agreeable jobs for deserving employees. The difference was that, for the most part, the gold in the rock stayed where it was.

Mining conditions at Cripple Creek were not easy, but part of the problem was that the skills needed to mine the gold had gone, although a few braggarts hung around the bars to assert otherwise. Near Victor, for example, was an operation called the Mary Nevin, nicknamed, with successive disillusionments, the Maybe Never and finally the Never Never. In the same cynical vein, Texasgulf became Texasgoof; Silver State came to be known as Silly State; the Fair Chance mine degenerated in reputation to the Fat Chance. The Mary Nevin was typical of dozens of small western mines in the 1970s and 1980s. Its tiny headframe could be seen half a mile below the road from Cripple Creek to Victor. Some mines elsewhere used slusher hoists bigger than the main hoist of the Mary Nevin.

In such a situation the promoters would raise money by selling stock. The proceeds went into their pockets as finders' fees, management fees, salaries, and expenses. Some went into assembling whatever equipment some dealer in tired iron managed to pass off onto them. They would dig a mining crew out of the bar. The hoariest veteran or the man with the loudest mouth would be appointed foreman. The blind led the blind for a while as they fought with worn-out and inappropriate equipment and other problems, natural and self-induced, until the money ran out. Then the promoters went back to promoting stock, and the miners returned to the bar to discuss their prowess.

Among the miners were one or two real professionals, men who had worked in mining and heavy construction from coast to coast, border to border, "and then some." Such men were few in number. If, of three such men, one was an ill-natured prima donna and another was an alcoholic, that did nothing to help matters.

If, in contrast, a man's sole experience of mining came from outfits like the Mary Nevin, broadened by barroom tales from New Mexico or Arizona, he might be inclined to strut and swagger and call himself a miner, while knowing almost nothing of the miner's craft and never having seen an efficiently run, full-sized mine.

In sharp contrast to the Mary Nevin were "rock factories," such as the Henderson molybdenum mine west of Denver, which took Amax ten years and $400 million to bring into production and which churned out a staggering 35,000 tons of ore per day. There a man could spend years driving a scooptram

and blasting drawpoints, after which he, too, would call himself a miner, although knowing little more of mining than
the operator of any similar piece of heavy equipment on the
surface.

Many such men lived around Cripple Creek—tough, coura-
geous, and hard-working but ignorant of mining, deceived by
their own eloquence, and intolerant of direction by outsiders.
The skillful all-round miners and mine foremen who worked
by the thousands all over Canada were either a rare breed south
of the border or they stayed away from Cripple Creek in great
numbers. Too many people at all levels in the hierarchy were
acting out a legend whose substance had leaked away, leaving
a husk of empty verbiage.

One mine at Cripple Creek ideally suited this propensity.
This mine shut down for the winter. Its crews worked no night
shifts, blasted no rock, mined no gold; in fact their work was to
talk about mining. The mine made a steady profit. This prodigy
was the first headframe a visitor passed on the main road into
Cripple Creek. It was called the Molly Kathleen—it did not
even have a sarcastic nickname. It was fitted up as an underground tourist attraction. Some of the men from the working
mines moonlighted there as tour guides. The Molly Kathleen
made a profit; the mines that struggled to produce gold did not.

John bought a house on Gold Hill. It was known locally as the
Cold House, and so it was. Sheltered by trees, it fronted onto a
red gravel road called Golden Avenue. In summer burros grazed
on the grass behind the house, as did the neighbors' horse. In
this way John started out as "the new mine foreman" at the
Ajax mine.

The Ajax had once been one of the bigger mines of the
Cripple Creek district. Opened around 1900, it had gradually
been sunk deeper with the passing decades until now its one
shaft went down from an altitude of 10,500 feet to a sump 3,400
feet below. Its latticework headframe on the southern slopes of
Battle Mountain overlooked the town of Victor. From town the
headframe could be seen silhouetted against the sky, while its
rock dumps spilled down the mountainside, partly burying a
jumble of building foundations and scrap metal, and blending
eastward with those of the Portland mine.

At 3,100 feet below the shaft collar, a drift connected the Ajax
shaft to the six-mile Carlton drainage tunnel, which had been
driven between 1937 and 1941 with the intention of draining

every mine in the district. The Carlton tunnel was the last and deepest of a series of drainage adits that penetrated the hills of Cripple Creek. Like some similar schemes elsewhere, it had done little more than attend the district's retirement. About thirty levels ran from the Ajax shaft above the Carlton tunnel, but they were entirely mined out. Current interest focused on the ground below the 3,000-foot level. Below the tunnel level a cutout had been made at 3,200 feet, while a shaft station and a short drift had been excavated at 3,350 feet. The shaft went below the 3,350-foot level for another 50 feet as a sump.

When the price of gold rose in the early 1970s, after being deregulated in 1968, the owners of the Ajax took fresh interest in the old mine, which, after years of disuse, was all but derelict. At that time the shaft was down only as far as the Carlton tunnel level. The lengthy process of reopening the shaft was begun, and in the early part of the decade, it was sunk a further 300 feet. In the mid-1970s the owners found a larger company to bring the mine into production on their behalf. The venture had its ups and downs. The shaft bottom was left to flood and then was pumped out again more than once. If abandoned, the shaft would flood only to the Carlton tunnel level because the water would then fill the shaft station ankle-deep before flowing out through the tunnel and so, ultimately, into Four-Mile Creek.

The operators of the joint venture began by installing themselves in a disused concrete mill building on the road between Victor and Cripple Creek. There they established offices with an ample supply of carpets, desks, filing cabinets, telephones, typewriters, potted plants, and all other modern and necessary conveniences. By 1981 the mill itself had been reconstructed, engineered on a lavish scale to treat the ore that was soon to flow from the mining operations on the hill. Thus comfortably protected from the elements, the company technical staff could produce plans, studies, and cash-flow projections without ever soiling their hands with the rude realities of getting rock out of the ground. By the time John appeared on the scene something like eight years and as many millions of dollars had been spent in doing little more than studying the situation. The time had come when this idyllic paralysis by analysis could no longer be sustained indefinitely, and they had to face the disagreeable business of going underground and breaking rock.

The nerve center of this operation was the general manager's office, where George H. Bertelsmann guided the ship with a firm hand. Bertelsmann was a grizzled veteran and looked every inch the part. A gold-painted hard hat adorned a bookshelf behind Bertelsmann's desk, while a poster on the wall exhorted the visitor to "Find gold!"

Accommodations at the mine were more basic. The architecture was in the best mining industry traditions of concrete and corrugated iron. A cavernous timber-framed warehouse was sheeted with iron that groaned and creaked in the mountain wind. Peering in through a crack or a dusty window pane, the visitor might see piles of machinery where rust vied for control with flaking paint and black grease. Reels of cable, boxes of electrical fittings, and piles of nondescript metal objects of uncertain function filled out the building's interior.

The mine dry was a low building, warm inside and smelling faintly of soap and socks, with muddy footprints across its concrete floor. Rows of metal lockers, their gray paint worn and scarred, filled one side of the building; shower and latrine cubicles occupied the other. From the roof hung clusters of mine clothing—oilskins, boots, hard hats, ragged jeans, wrenches, lamp belts, and self-rescuers—hoisted there on pulleys to dry while their owners were away from the mine. Here and there along the walls, hooks were randomly adorned with items of clothing, thick with oil and dirt, belonging to some forgotten miner who had quit so long ago that dust coated his possessions. Sunlight filtered through windows bleared with red dust thrown against them by the wind, which the rain seemed never to wash away.

The neighboring building was backed half into the hillside. From its front, taut hoist cables ran up over sheave wheels in the headframe. At the base of the headframe, within its slightly spraddled main legs, was the mouth of the Ajax shaft, exhaling wisps of damp mist.

The foremen's office occupied one room of the hoisthouse. A steel desk and a chair sufficed for the meager needs of the underground foremen, while a more ample installation, with a telephone and shelves of parts catalogs, surrounded the surface foreman's throne. Mine foremen came and went with the waxing and waning of work underground—no mining, no miners, no mine foremen. But as long as planners planned and engineers engineered and managers managed at the offices down the hill, the surface foreman remained secure. His more ample appointments announced the greater permanence of his position.

Next to the mine foremen's office was a room with racks of mine lamps on the walls and a decade-old safety poster framed behind cracked glass. Through another door of the lamproom was the hoistroom, centerpiece of which was the hoist, gleaming under electric light and glistening with oil and grease. Apart from the sound of someone hammering on a steel plate in the adjoining workshop, the first sign of life was the hoistman, half hidden behind his feet propped on the hoist controls, for such is the curious quietness that pervades the surface works of an underground mine.

This was the scene that greeted John as he went up the hill with Eric Bowman, the mine superintendent.

19

HEAVEN-HIGH, HELL-DEEP

You can see the Ajax mine in old photographs if you know where to look. The buildings of the larger and more famous Portland mine are more prominent, and the photograph will often be captioned accordingly, but in the background or to one side can be seen the furnace stack and shaft house of the Ajax mine.

Access to the underground workings was through a single shaft measuring 15′ × 6′ in rock. We must look at the arrangement of this shaft with a degree of detail that may at first sight appear tedious, but it plays a significant part in our story.

In contrast to the situation in Canada, enclosed headframes are not permitted in the United States because of the risk of fire. The old shaft house had long ago been torn down and replaced with a steel headframe, some ninety feet high, which had been bought from an abandoned coal mine. The steel lattice structure on the exposed hillside was open to the four winds of heaven.

The shaft was framed with timbers that divided it into three compartments. Skips ran in two compartments, while the third contained a ladderway, pipes, and cables. Men, rock, and materials were conveyed in two skips, each of which had a maximum capacity of one and a half tons of rock. Each skip had standing room for four men.

Most hardrock mines hoist and lower men and materials in cages and hoist rock in skips. In medium-sized or large mines this is done in separate shafts, or by independent hoists using separate compartments of the same shaft. At South Crofty, hoisting 900 tons per day, men and materials traveled in Robinson's shaft, while rock was hoisted in Cook's. At Dome, no. 3 and no. 6 shafts were divided into five compartments each, with a pair of five-ton skips, a forty-man cage, a cage counterweight, and a

CH

compartment for pipes, cables, and a manway. Considering that the Dome, hoisting 2,800 tons per day, was only a medium-sized mine, it can be seen that the hoisting works on the Ajax shaft were on a modest scale.

Men rode in the Ajax shaft by clambering into a skip bucket whose rim was about hip-high. A pair of removable steel-mesh gates confined their shoulders, and a roof protected them from objects that might fall from above. As they rode the skip, the shaft timbers zipped past their faces. When the skips were required to hoist rock, the gates and a locking pin were removed from each one so that they could be loaded and dumped without hindrance.

Communication with the hoistman was achieved by pulling on a wire clothesline that hung from pull-switches lag-screwed to the timbers every few hundred feet down the shaft. The pull switches rang a bell in the hoistroom. By making a coded series of pulls, the men in the skip or on a shaft station could indicate to the hoistman where they wanted the skip to go. Most mines have pull-switches or push-buttons that allow signaling from shaft stations only; shaft inspection crews use other methods of signaling to the hoistman when they are between levels. The continuous signal line down the Ajax shaft had originated in the rehabilitation work that had been going on intermittently for nearly ten years and in the need to signal from any point in the shaft.

The mine was owned by a company based in New York that had neither the capital nor the expertise to bring it back into production. They had, therefore, entered into a joint venture with another company capable of remedying those deficiencies. The relationship had been a stormy one, with the two parties sulking and pouting and spitting at each other from time to time. During periods of harmony the operating company made what progress it could with bringing the mine back into production, and indeed was under heavy pressure from the terms of the agreement to do so, no matter what obstacles stood in the way.

From time to time directors from New York visited the mine. They no doubt reckoned themselves fine fellows in the carpeted palaces of downtown Manhattan, but their stature was remarkably diminished in the raw world of western mining. John drew a sadistic amusement from their obvious fear and disgust at the wet filth of the Ajax shaft. Some of them found pressing engagements on the surface that deprived them of the pleasure of an underground visit.

As we have already noted, the bottom 300 feet of the Ajax shaft would, if abandoned, flood to the 3,100-foot level, where it was connected to the Carlton tunnel. After the shaft had been deepened in the mid-1970s, it had been left to flood, pumped out, and left to flood again at least once. When John arrived, the operating company was just beginning to pump out the bottom workings one more time.

Diamond drilling had indicated the presence of rich veins of gold ore still unmined in the granite at about the 3,100- and 3,350-foot levels. The company's plan was to pump out the shaft bottom, repair the skip-loading system, drift on these two levels, and then mine this ore at a rate of fifty tons per day. This was more easily said than done.

Above the Carlton tunnel the rock had been draining into the tunnel for forty years and was therefore dry to all intents and purposes. However, as soon as the company started to pump from below the tunnel level, water poured into the shaft like a tropical rainstorm from every crack and fissure in the rock. The water was charged with lime, which coated everything with a cementlike deposit. Metal corroded in the heat and wet, which, in addition, played havoc with electrical equipment.

Water was pumped from the shaft by means of a submersible electric pump hung on a cable from a winch. The pump discharged through a four-inch fiberglass pipe that gushed water into a ditch on the 3,100-foot level and so into the Carlton tunnel. As the water level sank, the pump was lowered on the winch, and twenty-foot lengths of discharge pipe were added one by one. The submersible pumps were, no doubt, a fine product. They suffered from a minor deficiency: they disliked being submerged. After a few days of submersion, water penetrated the seals and shorted the motor windings. The manufacturer's representative in Denver flatly stated that this was impossible—but was glad to supply additional pumps at a price. Three pumps were needed on hand or being repaired to keep one continuously at work. This did not accelerate the draining of the shaft; sometimes the water level stopped sinking and came back up again.

The company needed to enlarge the cutout at the 3,200-foot level. Therefore, as soon as a pump was sucking and guzzling and snoring at the water below that level, Jim Young and Gene Leaf went at it with jacklegs to do what was necessary. Jim tended to be generous in his use of dynamite, with the result that flying rock cascaded into the shaft, smashed the shaft timbers, broke the fiberglass discharge pipe, and shredded the insulation on the 440-volt power cable to the pumps. When blasting granite in a confined space, damage of this nature was almost inevitable, but that did not make it any easier to repair.

The next crew to go down the shaft would find the pump stopped, the power cable kicking and flaring and sparking in the darkness of the shaft, and the water rising. The cable was never entirely repaired, and as the shaft was soaking wet, almost everything around it became live with stray electricity. When people were working around the open shaft, it was a little disconcerting to get a mild electric shock from every handhold. John acquired a habit of touching anything lightly before grasping it. This became automatic, even away from the mine, and persisted for years afterward. People thought it strange that John would touch them lightly before shaking hands with them, but he was just checking them for stray currents.

As work began on levels other than the 3,100, telephones were needed. One system, consisting of two U.S. Navy surplus telephones, connected the 3,100 with the hoistroom; it worked well. The other system consisted of telephones made specifically for mine use. This system had one little problem. These telephones disliked damp; most mine atmospheres are damp. They also depended on batteries, which leaked their charges in damp conditions, with the result that these telephones were often found to be dead when they were most needed. Communication was therefore unreliable.

Even the signal cables hanging down the shaft added their quota of mischief. A single length of cable hung from each pull-switch, its bottom end tied off to the next switch. As time passed, the cable stretched under its own weight with the result that a loop of cable hung down from the tie-off. Not infrequently a skip would catch the loop and tear it loose. Either the skip would come to the surface festooned with cable, or the hoistman would hear "one bell" as the cable wrenched itself free of the pull-switch. One bell meant "stop the skip"; one bell rung while the skip was traveling at full speed generally meant some dire emergency. As men's lives depended on the hoistman's exact compliance with whatever signal he received from the shaft, he would do precisely that. In either event, the resulting confusion took hours to unravel. Sometimes a loop of cable would come aboard the skip while men were riding in it. John would never forget the look on Bill Wood's face as he fought off a loop of cable that had fallen around his neck with the obvious intention of garroting him.

After several months of wearisome and frustrating labor, the steamy darkness of the 3,350 shaft station was revealed once more to view. The shaft bottom was full of rock and debris to within a few feet of the station sill. Two men had the unenviable task of working knee-deep in water to shovel rock into the skips, which were thus unavailable for any other use. Meanwhile water was pouring down onto them from all directions. It was almost impossible to hear what anyone said because of the roar of falling water and the thrumming of the pump in the confined space. Such work was hard to endure. It was fortunate that the water was warm, because work in the Ajax shaft left a man wondering if he had not turned into a fish.

Supposedly, once the 3,350-foot level was drained and rockwork could begin on this and the 3,100, it should be easy enough to hoist the required daily fifty tons in a shift. If each skip held one and a half tons and took fifteen minutes to reach the surface, then in theory six tons could be hoisted in an hour, forty-eight tons in eight hours or a bit longer, allowing for the beginning and end of the shift and a lunch break in the middle. There were snags. Many of the snags were small in themselves, but an apparently infinite number, some of them interlocking with each other, could at times bring the whole operation to a halt or, at least, waste precious hours that could in no way be recovered. Not only was Murphy's Law in effect; it seemed that Murphy had designed the whole operation for his amusement.

The method of loading the skips was primitive in the extreme and very slow.

Back in Cornwall in the 1850s, someone came up with the bright idea that the way to fill a skip was not to load it shovelful by shovelful but to equip the loading station with a box the size of a skip, filled from a chute. When a skip arrived at the loading pocket, the skiptender could release an exact load of rock into the skip and immediately signal to the hoistman to take it away. Where twin skips ran in parallel shaft compartments, the descending empty skip partly counterbalanced the ascending full one. While a full skip was on its way to the surface, an empty one was already on its way down to where the skiptender was filling the other loading box. It worked like a charm.

This system was first used at South Wheal Frances, near the somber village of Carnkie, where clouds and rain so often swirled around the engine stacks of the mines surrounding the village. The hoisting plant at South Wheal Frances was hailed as a masterpiece of contemporary mining practice. This same system, although improved with steel chutes operated by compressed air, enabled the Dome mine to hoist nearly 3,000 tons of rock in a pair of five-ton skips between 3:00 P.M. and 2:00 A.M. every night as reliably as clockwork.

At the Ajax shaft in its palmy days, ore had been hoisted by rolling loaded mine cars onto a two-deck cage underground at each level and rolling them off again at the surface. That system was, and still is, a common alternative to skip hoisting, especially at coal mines, where the intention is to avoid breaking up the coal more than necessary. At Rammelsberg 950 tons per day were hoisted in the Rammelsberg shaft by this means, and the mechanized system for loading, moving, and unloading the cars was marvelous to behold.

The current system at the Ajax shaft was absurd, neither fish nor fowl. A skip was lowered to the loading station. A slusher hoist was used to scrape rock into a chute and thus into the skip. Two men spent ten or fifteen minutes loading each skip, and for every one and a half tons that went into the skip, 50–100 pounds of rock went down the shaft.

Eventually the skips would land on broken rock before they could reach the lip of the loading chute just below the 3,350-foot level. Some unfortunates would then have to face the ordeal of mucking the shaft bottom, often for several shifts, before rock hoisting could be resumed. Only a few more days would pass before the problem repeated itself.

Compacted mud and rock stuck in the skips, thus reducing their payload. This problem, like that of spillage from rock hoisting, is as old as mining. A properly designed hoisting system takes this into account and incorporates an arrangement at the shaft bottom for collecting and disposing of spillage, and another in the skip dump for cleaning the skips. Cleaning skips and emptying the spillage pocket is routine; no one thinks anything of it. At the Ajax shaft things were different.

At the shaft bottom at least it was warm. The skip dump, however, was forty feet up in the open headframe and neither well designed nor solidly built. On a winter night skips came up from the steamy depths of the mine leaking wet slime, which formed a deposit on the skip dump and froze. Eventually a skip would jam in the dump. A man had to crawl around the headframe in temperatures of ten and twenty degrees below zero, with the wind tugging and pushing at him, to try to pry it loose. From time to time he had to retreat to the hoistroom to thaw out, and the whole operation could take five or six hours. Meanwhile the skip loading crew sat around underground with nothing to do.

The hoisting system was threatened by worse possibilities than hung-up skips. The hoist had two drums, each driving one skip. The drums could be clutched together so that each skip partly counterbalanced the other, or they could be unclutched to allow each skip to run separately. After running the hoist unclutched, the hoistman had to check the position of each skip when he wanted to clutch the drums together again. Interesting things could happen if he did not.

One midday John was sitting on the edge of the rock dump eating his lunch and admiring the view, which was spectacular. At his feet lay Victor, so close below that he could watch the inhabitants going about their business. Behind Victor was the sparsely wooded bulk of Straub Mountain, then range upon range of hills spreading far away to the sawtooth outline of the Sangre de Cristo Mountains. John could gladly admire this view for as long as his duties allowed; weather permitting, he would take his lunchpail and sit on a plank at the edge of the dump that sprawled down into the outskirts of Victor.

This time he turned his head at the sound of the hoist and looked at his watch, thinking that the skip loading crew were unusually eager to get back to work. He saw one of the skips move slowly up into the dump where the skip bucket overturned as it was supposed to. His jaw dropped open as the skip continued its upward path. The bucket swung upside down. Some rocks and debris that had been stuck inside fell slowly down the shaft, followed by two lengths of timber guide rails that the skip dislodged. The skip grated to a halt.

John and Eric went down in the other skip to inspect the shaft for damage and to retrieve the guide rails. According to the skip loaders, no debris ever reached the shaft bottom; it had lodged in the timbers or manway compartment, or had bounced out onto a shaft station somewhere on the way. The biggest piece that they found remaining from two twenty-foot lengths of 4″ × 4″ hardwood was a splinter six feet long. It took ten hours to get the skips running again. John felt that he might as well flap his arms and try to fly to the moon, which glared down through the clear mountain sky at night, as try to get fifty tons of rock hoisted in eight hours.

One man with a jackleg can break more than fifty tons of rock in a shift, so as soon as anyone did any serious drilling and blasting, the whole operation became "muck-bound." In other words, no one could move any rock because there was nowhere to put it.

Part of the trouble was that the joint venture agreement required the mine to be "in production" by a certain date. Rehabilitating old mine workings can be a bottomless pit of expense, and it is difficult to predict how long such work may take. The date arrived, and Bertelsmann announced that they were "in production" from henceforth, a pronouncement tantamount to moving into a half-built house and then saying that one expects to live there.

"Production" required numerous machines—jacklegs, mucking machines, slusher hoists, fans, and pumps—all powered by compressed air. The compressed air was supplied by a well-worn electric compressor in a shack on the surface. This machine had sufficed for rehabilitation and minor exploration work but was in no way adequate for the demands now placed upon it.

When the mine was fully at work, the frustrated miners would often find their machines running with less and less enthusiasm. The hoist brakes also depended on compressed air pressure, and when the pressure became too low for their safe operation, the hoistman would turn off the valve on the supply pipe going down the shaft. The roaring and burping of the machines would subside to a futile chuffing and hissing, and finally to a fading sigh. The miners would have no option but to sit down and wait for pressure to build up again in the system.

John, as the shift foreman, was the natural recipient of much of their frustration. Goaded to sarcasm, he told more than one aggrieved voice on the telephone that if he, John, could have dreamed up another compressor out of thin air, he would long ago have done so; however, as the speaker obviously knew something that John did not, he was welcome to try. This kind of thing often seemed to happen on night shift, when John's only means of relieving his feelings was to write in his shift report: "Compressed air NFG—again."

The gentle reader may well ask why these things were not fitted up properly. The joint venture agreement was driving the company's vice-president responsible for the operation. The veep was driving Bertelsmann. Bertelsmann was driving the unfortunate mine superintendent like a horse. If John ever asked why certain things could not be done, Eric's eyeballs would rotate 360 degrees in azimuth and elevation—separately—and one of three answers would pop out. "We don't have time." "It's not in the budget." Or "The head office back east would have to do a feasibility study, and approval would not be forthcoming for at least six months."

Things might have been easier if John had always had a full crew. Especially as they struggled through the winter, absenteeism sometimes reached 50 percent. Those who did come to work had to be assigned where they were most needed, and they were understandably annoyed at being shuffled from workplace to workplace.

Pneumonia has always been a danger in the thin air of the high-altitude western mining camps. If a man worked his shift in sweltering heat at the bottom of the Ajax shaft and came to the surface into a winter night at ten or fifteen degrees below zero or into a blizzard, it was not surprising that coughs and colds flared up into pleurisy or pneumonia. On many a cold night John would come to the surface in his sodden clothes and wet, greasy oilskins. Even in the time it took to open the shaft gates, clamber out of the skip, slam the gates, ring the release signal, and sprint sixty feet to the heated hoistroom, his clothes would be stittening on him as he burst in through the hoistroom door.

Most houses in Victor and Cripple Creek dated back to the 1890s and early 1900s. When frost struck deep into the ground with little snow cover, the small-bore water pipes, buried at shallow depth, froze and burst. A man often had to stay at home and rescue his property.

The standard method of thawing pipes was to attach one lead

from a welder to a heavy-gauge copper "thaw wire" that supposedly was attached to the water line serving each property,
which surfaced somewhere in front of each house. The other
lead was attached to pipework inside the house, and the welder
was turned on. The heavy current passing through the pipe
would melt the ice in fifteen or twenty minutes, after which the
homeowner, if wise, would leave a tap running for the rest of
the winter. It was not always possible to tell where the electric
current would go. One man who owned a welder refused to use
it for thawing pipes; he had tried it in Victor and set fire to six
houses at the same time.

At 10,000 feet the camp was above the worst of the snow-
storms that blanketed the country around Colorado Springs and
Canon City, but when it snowed at Cripple Creek, it did so in
no uncertain fashion. Men who lived in outlying parts of town
were snowed in. Besides, with conditions at the mine being
what they were, John was often reluctant to go to work and was
not surprised that his crew felt the same way.

Especially on the night shift, there were other absences not
entirely unconnected with a watering hole called Zeke's. And
then there was Meatball. Most large men around mining camps
tend to be of a gentle disposition. Meatball was not.

One Monday morning Meatball met John with the words,
"Diggy Brick won't be in today."

"Why not?" asked John.

" 'Cos I beat the shit out of him," replied Meatball, grinning
from ear to ear.

Diggy Brick was a tramp miner whose main tramping ground
had been the Coeur d'Alene district in northern Idaho, but who
had for the time being taken up his abode in a shack in the
outlying parts of Victor. The voice that emerged from his gan-
gling frame addressed the world in profane terms delivered
wearily in a dismal undertone.

It had been a bleak Sunday afternoon. Patches of snow filled
the hollows in the hills and lay in drifts among the derelict headframes and yellowish-brown rock dumps overlooking Cripple
Creek. There was no snow downtown, except for a few patches
in the shelter of ruined buildings, but the air was hard and cold
under a leaden sky. The main street swooped in a catenary from
one side of town across a shallow gulch and up onto Gold Hill.
The city of Victor scaled the airy heights of civic pride in its
latter years the day they installed a new trash can at the inter-

section of Third and Main, but Cripple Creek had a stoplight at the bottom of town.

Two pickups drove toward the stoplight from opposite sides of town along a main street cleared empty by the bitter mountain wind. One pickup was driven by Diggy Brick, the other by Meatball. The two men were not the best of friends; the mine and the empty winter mountains did nothing to soften such antipathies. As they passed each other, Diggy Brick stopped in the middle of the intersection and rolled his window down. Meatball did likewise. Diggy Brick unburdened himself of the simple inquiry, "Why don't you go fuck yourself?" Diggy Brick's natural pace was a languid one, not nearly quick enough to escape Meatball, who sprang from his truck, grabbed him, and cleaned his clock in no uncertain style. Consequently on Monday morning Diggy Brick stayed at home in great agony of soul to nurse his face and feelings.

The permanent population of Cripple Creek was about 600 people, that of Victor 400; consequently the two towns supported only a limited number of bars. Meatball frequented those bars, as did most other people who worked at the mine. With the passage of time came a steady toll of employees who were off work for a day or so following impact with hard objects. Those hard objects tended to be in and around bars frequented by Meatball. When Meatball remarked one day that a coal mine might be opening up near Canon City, John grinned innocently and agreed that it sounded like a good proposition, hoping that Meatball would quit and go there—or indeed anywhere other than Cripple Creek. With absenteeism from so many other causes, the company could not afford to have Meatball going around damaging people. Yet it was not the company's business what a man did in his spare time, and Meatball was a half-decent miner where such talent was in short supply. The problem solved itself.

One spring afternoon John was gathering his gear to go on the night shift. The sun was just coming out after a sharp thunderstorm; John wondered if he should not telephone Eric to say that he could not go to work because lightning had incinerated his work clothes as they hung on the line to dry. You could believe anything around Cripple Creek, no matter how bizarre, and it would probably be true. Nevertheless the mine foreman was not supposed to pull stunts like that, so John sighed and continued his preparations to go to work.

Someone knocked at the door. On the doorstep stood Meat-ball with one hand splinted and bandaged and one eye covered with a huge, bulbous plaster.

"Guess I won't be coming to work tonight," said Meatball, grinning.

"What on earth happened to you?" replied John to this self-evident truth.

"Other guy knew karate." The whole sequence of causes and effects was so familiar that only the fewest and sparsest exchanges sufficed to fill in any untoward details.

"But I got hold of him in the end," Meatball remarked with an air of finality. "He's not one of our guys," he added in response to the look on John's face.

John left the district soon afterward, and the problem was no longer his concern. At about that time, too, the problem of Diggy Brick solved itself for the balance of John's tenure.

The problem of Diggy Brick was that he was an accident looking for a place to happen. It seemed that every loose rock in the mine converged on him. "Diggy Brick got slabbed" was a message repeated with monotonous regularity. In the event of a temporary paucity of loose rocks, he managed to injure himself in other ways by having various parts of his anatomy in the wrong place at the wrong time. John and Eric suspected that he was a drug addict, but as they knew nothing of the symptoms, they could do nothing about it. Toward the end of John's time at Cripple Creek, their suspicions were amply confirmed.

The road between Cripple Creek and Victor followed a level route contouring around the hills and gulches. The road swung in sharp bends around the noses of spurs projecting from the central hills of the camp. One afternoon Diggy Brick missed a bend in his pickup and went careening down a boulder-strewn slope. The pickup disintegrated on the way down, and the pieces came to rest, with Diggy Brick entangled in the wreck-age, about a thousand feet down the hill. As his rescuers carried him off to the hospital, they found that he was so heavily drugged that his heartbeat had slowed down. That was why he had driven off the road; it was also why he had not bled to death.

With so many difficulties to contend with, the mine was appallingly inefficient. There was more to the story, however.

John's opposite number, the foreman on the opposite shift, was a man named Jack Eilenbarger. Eilenbarger's obscure antecedents became clearer as time went on. He was a big, rangy man in his forties, extravagant in speech and gesture. To be sure, he looked the part of the seasoned mine foreman. He talked of having been the superintendent on a shaft-sinking job in Wyoming, of having worked on the construction of the Eisenhower Tunnel, and of mining around Idaho Springs, a gold-mining area west of Denver. It all fitted, and the picture had added up well enough to convince the operating company to hire him as a shift foreman. When he and John met on occasion, good old Jack was all smiles and geniality. John noticed, however, that Eilenbarger slandered the other mine staff as fools and suspected that he was getting the same treatment behind his back. By keeping his eyes and ears open and his mouth shut, John discovered that this was indeed true. He was amused by the crew's attempts to play the two of them off against each other.

John noticed that Eilenbarger's crew was singularly unproductive, especially on the night shift. On one occasion he and Eilenbarger went underground together between shifts. If two or more men rode the skip, the man closest to the signal cable rang the signals. The code was not hard to learn and was almost standard throughout North America. But Eilenbarger asked John to ring the signals and never once touched the cable. Part of the foreman's job was to travel in the shaft frequently and often alone. Most people knew the skip signals as well as they knew their own telephone numbers. If Eilenbarger had been at the mine for six months and still did not know the skip signals, the implication was such as to cause John to raise his eyebrows far enough to push his hard hat off his head.

When he found a Louis L'Amour novel in the desk that he shared with Eilenbarger, the picture became clearer still. Eilenbarger spent his time cruising around on the surface in the company pickup or reading Westerns in the foremen's office. His crew wore out the lunch bench and talked about mining. And little or nothing got done.

One day a visitor passed through, an elderly mining engineer, the type of semiretired consultant whose experience is long and varied. John happened to be in the mine foremen's office, and the man paused to pass the time of day. Eilenbarger's hard hat lay on the desk with his name on it in Dymo tape. The man asked John, "Is that Jack Eilenbarger?"

"Sure," replied John, "He's my cross-shift."

"So Jack made it to shiftboss, did he?" The man chuckled and shook his head.

"Well, he was a shaft superintendent in Wyoming, wasn't he?" retorted John. Whatever he might suspect, he was not going to let an outsider demolish Eilenbarger's reputation without putting up some sort of resistance.

A pregnant silence followed. John and the visitor eyed each other as if to ascertain who was kidding who. The man shook his head, chuckled, and walked away.

These problems, although trying to the patience, at least comprised such familiar ingredients as Murphy's Law and the vagaries of human nature. Mining at Cripple Creek did, however, have certain features above and beyond what mining people are normally expected to put up with. Those who worked in the Ajax shaft soon found that they walked in the valley of the shadow.

GAS AND GHOSTS

One of the less endearing features of the mines at Cripple Creek was the fact that the rock emitted carbon dioxide gas. The ore was in veins that ran through the throat of an ancient volcano. The very last of this volcanic activity, dead for tens of millions of years, was the warmth of the rock and the seepage of carbon dioxide.

A drill hole in the wall of the 3,100 shaft station blew gas steadily and under a pressure that John could not confine with his hand. Stupidly he sniffed at the borehole. So sudden and complete was his body's refusal to inhale whatever was coming out that he involuntarily pulled his head back as though struck in the face. For the most part the gas seeped from cracks in the rock and collected in low places in the mine, as carbon dioxide is denser than air.

The method of guarding against asphyxiation was the Wolff flame lamp. The gauze-covered safety lantern was originally invented as a safe source of light in coal mines where methane came out of the rock and continues in use today to detect methane or lack of oxygen. The flame of the Wolff lamp requires a substantially higher percentage of oxygen in the air to burn than the human body needs for survival. If the flame lamp goes out, it therefore gives warning of an oxygen-deficient atmosphere with a useful margin of safety.

Each crew and each foreman carried a lighted flame lamp in addition to the electric hat lamp that each man wore for illumination. The flame lamps burn well in mines whose deepest workings are at or below sea level, where the air is dense. The Ajax shaft collar was at an elevation of 10,500 feet, and therefore the 3,350-foot level was over 7,000 feet above sea level. As a re-

sult the lamps did not burn well and were easily extinguished by being struck or dropped, or by the all-pervading water. Once extinguished, they were not easily relit. Without a lighted flame lamp to warn him, a man could move into a pocket of gas without knowing it and be rendered unconscious before he could escape.

The warm rock in the mine heated the air, which rose up the Ajax shaft and drew fresh air in through the Carlton tunnel. As a result, the workings at and above the tunnel level were mostly self-ventilating. Even so, John sometimes walked into old workings where the air was still. By letting the lamp hang from his fingers it would swing along at the level of his shins as he walked. In quiet, ill-ventilated drifts, he would see the flame weaken in the layer of invisible gas along the floor. If he chose to continue, the flame would go out. At that point prudence dictated that he turn about and retreat.

The problem lay in forcing air continuously into the workings below the 3,100-foot level. This was accomplished by an electric fan, two feet in diameter, hung on the 3,100 shaft station. The fan drew fresh air from the stream of air flowing in from the Carlton tunnel and forced it down the shaft through fiberglass ducting. After it had done its work of ventilating the bottom workings, this air joined the air currents flowing up the Ajax shaft to the surface.

Once, when the fan had been off for some time, John lowered a flame lamp on a rope down the shaft from the 3,100 station. The lamp went out abruptly less than six feet below the shaft station sill. The shaft was full of gas. When they restarted the fan, they stood back from the shaft while the gas joined the natural airstream to the surface. They left a flame lamp standing on the floor a few feet from the shaft. An ankle-deep layer of white vapor crept out from the shaft like a tide, extinguished the flame lamp, then ebbed once more as the shaft cleared of gas. Carbon dioxide is said to be invisible; no explanation of this phenomenon was forthcoming.

A procedure was laid down for entering the mine, to be followed by every shift crew before going underground. Two flame lamps were hung in a skip at the surface. The skip was sent, empty, to the 3,100 level, then brought back to the surface. If the lamps were still alight, the crew could safely go underground as far as that level. Once there, they had to repeat the procedure for the bottom part of the shaft. Only then could they begin their day's work.

The idea will have begun to dawn on the gentle reader that the Ajax shaft had a quality of extraordinary malevolence. It seemed that the mountain resented men's intrusion into its depths and sought to kill them with an extremity of vicious cunning. The mountain had killed men before, and the mine was haunted by the spirits of the unquiet dead.

Most mines are haunted; it is only a question of degree. The scoffers have their answers, and sometimes they may be right, but a residue of experiences remains, involving the most sober and matter-of-fact of men, which defies rational explanation.

The story was told in South Crofty that in the 1950s an Italian miner turned the corner of a drift to find a bearded stranger sitting on a timber. The man wore a resin-cloth hard hat with a candle stuck to it with a dab of clay, and the long canvas jacket of a nineteenth-century Cornish mine captain. Such garb, and the candle on the hat, had gone out of use decades before. The Italian not only went to the surface but returned to his native land forthwith.

A friend of John's in Cornwall had a recurring dream all through his life, whose location he could neither recognize nor understand, which was a mystery to him. He never worked as a miner, nor did he ever intend to do so, until forced into it by circumstances in his forties. When he started work at South Crofty, he recognized the place in the dream, not only as the workings of a mine, but as a specific place in South Crofty on the 335-fathom level. The dream ceased.

Men left Wheal Jane, another mine in Cornwall, because of what they saw and heard underground for which there was no reasonable explanation.

In one part of the Dome—1832 drift, to be precise—John always had an overpowering and disagreeable sensation of being followed. In another part of the mine, in the early hours of one morning, he was climbing down the ladders through a raise from one level to another in a deserted area where he had gone to check for compressed air leaks. The mine was, in any case, sparsely populated on the night shift. He heard the rasp of nailed boots on the steel rungs of the ladders above him and heard the pipes in the raise knocking together as the ladders flexed under the other man's tread. There was no other man.

After the Ajax 3,350 level was drained, much work had to be done in the way of pipe fitting, hanging cable, installing ventilation ducting, and the like. John and his crew often had to work in the narrow confines of the shaft with the open shaft above them. When something falls down a mine shaft, it seldom falls straight down. More often it bounces from side to side. If it is of any size, it hits the bottom like a bomb and kills anyone unfortunate enough to be there at the time. Not infrequently when they were working in the shaft, unable to escape, they heard the heavy thudding of an object falling down the shaft above them. Tools and pipe fittings flew hither and yon and fell into the flooded sump as they flattened themselves against the shaft walls in terror, praying that the falling object would lodge in the ladder compartment before it hit them. Nothing ever reached them. Although they searched the shaft for whatever it was—and it was no small object—no one ever found anything.

In the 1970s, while the shaft was being deepened below the 3,100 level, the crews became lax in checking for gas before going underground. Two men signaled to the hoistman to lower them directly to the shaft bottom. The hoistman, being as lax as the miners, complied. As the skip approached the shaft bottom, the hoistman received a vague signal that he did not understand and stopped the skip. The bottom of the shaft had filled with gas. One man was later found dead in the skip; the other had climbed out and had fallen down the shaft. John reckoned it was that man's body that they so often heard falling.

John was riding the skip to the surface one day with Gerry Breitenfeld, the mine electrician. They heard a strange cry—apparently human—from a level that had been abandoned long ago and was supposedly deserted. They were riding the skip with their backs to the shaft station. Both heard the same cry. After a brief conference, both went back to investigate. There was no one there.

At other times John heard blasting when there was no blasting. Once he heard one of the miners call his name—the man's voice and intonation were unmistakable—when that man was in a different part of the mine. By this time John had been working underground for nearly ten years and was not likely to be deceived by the noises of the mine.

One night John went down to the 3,350, as he did at least once a shift, to visit Jim Young, Gene Leaf, and Charlie Snare, who were drifting on that level. The whole level was only 250 feet long, but the face could not be seen from the shaft because of a bend in the drift. There was no permanent lighting, and in the steamy darkness and pouring water around the shaft, a man was visible chiefly by his hat lamp. John had just climbed out of the skip and was latching the gate before giving a release signal to the hoistman.

Over his shoulder he saw one of the men come toward him from the face and pass behind him to go around to the back side of the shaft. He saw only the man's hat lamp but thought nothing of it. As they were a merry and garrulous crew down there, he was surprised that the man gave him no greeting or sign of recognition—no cry of "Hiya, pard!" yelled through the rush of falling water, the hiss of leaking compressed air, and the thrumming of pumps, enlarged in the echoing shaft station. Thinking that the man had gone to telephone the hoistman, John followed him. There was no one there. Moreover, all three men were at the face, hard at work amid the fog and ground-cracking roar of their Gardner-Denver 83s, punching another round into the red granite. John never mentioned the episode. Everyone knew that the mine was haunted; no one wanted to be reminded of the fact.

John and three of his men came close to joining those specters who haunted the mine. He had no premonition, and the event happened very quickly. Four of them were riding a skip down to the 3,350-foot level. At that time the water stood just six inches below the shaft station sill and was twenty or thirty feet deep. The hoistman knew this and would ease the skip down the last twenty feet, waiting for the "stop" signal from the men in it. Even if no signal came, he would stop the skip at the level. The position of each skip in the shaft was displayed in the hoistroom by a pointer on a circular dial. For exact positioning, the hoistman aligned marks painted on the drum of the hoist with an index mark on its frame. By this means, he could spot skips at each level to within inches.

Normally the skip would slide gently down to the 3,350 station and stop just as it touched the water. This time, where the shaft walls widened out to form the shaft station roof, John noticed with alarm that the skip was speeding up instead of slowing down. Faster and faster, they fell through the shaft station to hit the water with a hefty splash. Such was the speed of events that they were helpless. Jammed in as they were, the skip roof would come down on top of them, even if they did try to escape. When John felt the water swilling about his stomach, he realized that he was about to die. A picture flashed through his mind of gray water covering his face, lit from within by his hat lamp. Mentally he could already feel water splashing over his face. He recalled that someone who had nearly drowned had told him that the experience was not particularly unpleasant. The skip stopped abruptly; the water was up to their chests. They scrambled out onto the shaft station floor, which was level with their faces. They counted heads, streaming water and shaking with terror. The shaft station rang with John's oaths of anger and alarm.

The skip was fitted with "safety dogs." The "dogs" are a safety device used on shaft conveyances; they are set to bite the timber guide rails and bring the conveyance to a halt if the tension of the hoist cable on its attachment is for some reason released. The skip had hit the water so hard that the dogs had bitten and stopped it. Had it entered the water less violently, the dogs would not have bitten, and the men would have drowned. A loop of hoist cable lay on the skip roof. The shaft station was equipped with a telephone, but being of the kind purposely made for mine use, it was not working that day. Other than by climbing 250 feet of cramped and encumbered ladders in their waterlogged clothes to the 3,100 level, where the U.S. Navy telephone was, they had no direct means of finding out what had gone wrong.

Tentatively, John pulled the signal cable and gave the signal "hoist slowly." To his surprise, the cable began to move and straightened out without kinking. The dogs disengaged as the tension came onto the cable attachment, and the skip rose, spouting water from every crack and drainhole. John rang one bell and the skip stopped. The four men looked at each other doubtfully. Without a word they climbed into the skip, in which the water was still shin-deep. John rang the signal for the 3,100 level. In silence they rode the skip; in silence they disembarked on the 3,100.

John headed for the telephone and twirled the handle. His outraged inquiry could have been heard all over the mine. The answer was that nothing was wrong with the hoist. A trainee hoistman had been at the controls, with the regular hoistman looking over his shoulder. As the indicator showed that the skip was approaching the 3,350 level, the hoistman should have applied the brakes and cut off the power. Instead the rookie had cut the power but had then frozen at the controls, with the distressing consequences just described. After John had ascertained that the rookie hoistman had been sent home for the day, he and his crew, having rested their shaken nerves, returned to the 3,350 level to finish their work. At the end of a shift in the Ajax shaft, everyone came to the surface soaking wet anyway, so their sodden condition and the pools of water that formed about them on the floor of the dry aroused no special comment.

As he drove away from the mine, John met Bertelsmann driving toward it. "You know," said Bertelsmann, after John had described briefly what had happened, "it doesn't do to get too excited about these things." John drove home over the sunlit hills, shaking his head.

The nightmare of the Ajax shaft continued. John and Eric were riding down in a skip one day, moving slowly from the 3,200 cutout to the 3,350 level. The skip hung up on a rock that had lodged on a shaft timber. A hangup of this nature is especially dangerous because, if the skip frees itself before the safety dogs bite, the shock may be enough to break the cable. The cable will tend to break at the point of greatest strain, which is either at the sheave wheels in the headframe or at the hoist. The resistance of the cable trailing above the falling skip may keep sufficient tension on its attachment to the skip to prevent the dogs from biting. Even if the dogs do bite, 3,000 feet of 1-inch or 1½-inch steel cable falling down a shaft can be remarkably destructive. John knew this. Until that instant when the skip hung up and freed itself, he never knew that his body could produce so much sweat in so short a space of time. So rich and salt was the sweat of deadly terror on his upper lip that he wiped his nose on the back of his hand to see if it was blood. The cable held, and they continued on their way.

One hoist cable had been in service for many a long year—so long, in fact, that it had worn down smaller than the proper diameter. Each hoist drum carried three layers of cable when the skip was at the surface and nearly all of the cable was on the drum. If the cable was of the correct diameter, each layer would track smoothly onto the one beneath it as the hoist reeled it in. With the cable as worn as it was, it would jump when the ascending skip was about 300 feet from the surface as the top layer started to form on the drum. The skip would jolt, with a clang, to the alarm of anyone riding in it.

John could never quite see the funny side of this one. Many a time, he would be riding to the surface alone. The skip would jolt, and he would swear to dump his gear on the floor and quit as soon as he reached the surface. After another hundred feet he would reflect that, as he was the shift foreman, it would be irresponsible to quit in the middle of the shift; he would stay until the end of the shift and then quit. By the time the skip reached the surface, he would have reflected that he did not have another job in sight and had better stay until he had.

John had not been at Cripple Creek for more than a few months before deciding that he should start thinking about other employment. His experiences as time went on did nothing to erase that idea.

The project had come into being in its most recent form when the price of gold was bucking $700 an ounce and experts were predicting prices of $1,000 an ounce, or even $2,000. Most gold mines in production at that time were turning out bullion at a cost of $200 to $400 an ounce. However, since the end of 1980, the price of gold had been sliding steadily downward, and the gilt had come off the gingerbread for places like Cripple Creek.

The workings of the Ajax, and the Cresson decline over the hill, were probing into what was supposedly gold ore. Only a small proportion of the gold that is mined is visible to the naked eye. Nevertheless most hardrock gold ores have at least some assemblage of clues that allows the trained eye to distinguish between ore and waste. One of John's miners had Cripple Creek ore figured out in a nutshell. He came to John one day with big, round eyes after looking at a freshly blasted face, with the words, "There's yer arn an' yer floride an' yer green delirium, an', pard, that's *paydirt!*" Brassy iron pyrite, purple fluorite, and a green mineral that was supposed to contain tellurium were the concomitants of gold at Cripple Creek. Such showings were rare, and to John, it was clear that they were mining red granite, and not much of that. The assay results of sampling showed that this was regrettably true.

Whatever the company did, it turned to something that certainly was not gold. All of a sudden, there was a big panic to reopen the 3,000-foot level and extend an old drift a few hundred feet along a vein to where the geologists thought there was ore. This was easier said than done. A crosscut ran from the Ajax shaft across several veins that had been mined in times past. The vein that they were supposed to follow was about 1,000 feet from the shaft. The end of the drift that they were to extend was 1,000 feet from the crosscut. In the drift face was a couple of fractures that the geologists thought would widen out into a full-sized vein.

The level was already equipped with track, pipes, and ventilation ducting, dating from the 1950s or earlier, but they were so small and so badly rusted that they were unusable and had to be torn out. As the lengths of track and pipe were rusted together, this meant burning them apart with a cutting torch. Next, all these services had to be reinstalled. That was when they discovered that the brow of the 3,000 shaft station was too low to allow thirty-foot lengths of rail to be swung from the vertical position in which they were lowered down the shaft into a horizontal position for transportation along the level. The company had obtained a stock of forty-pound rail from somewhere, and it was all in thirty-foot lengths. That resulted in another panic to find someone who could cut thirty-foot rails into fifteen-foot lengths and punch the holes for fishplate bolts. It was experienced rail. The foundry stamp on each length read, "Joliet, Illinois, 1893."

To drive the 3,000-foot level, when all was ready, they had Gardner Denver model 83 jacklegs and an Eimco 12B mucking machine. To haul the muck from the face, they had a single one-ton end-dump muck car of the kind that was in common use before 1900 and is now seen in museums or as monuments in public places—and no locomotive. A locomotive and muck cars were not in the budget. Even if they had been, there was no time (as well as nothing in the budget) to construct a car dump, nor had anyone given thought to how they would get the ore into the shaft hoisting system. For the time being, two men took two shifts to muck each round. They pushed each ton of rock, carload by carload, 1,200 feet to a hole in the floor between the rails and dumped it into the top of a mined-out stope.

They drove the drift, following fractures in the granite, to where the geologists said that the ore was. There was nothing there.

It was the same with the Battle Mountain tunnel—a futile attempt to reopen an old adit in search of ore. The attempt bogged down within 200 feet of the portal while mucking out and retimbering a stretch of caved ground. Another crew was rehabilitating the six-mile Carlton tunnel.

Time went on and money was spent, all to no effect. Wherever they went, there was no ore; if there was, they could not get at it. Two stopes were started on the 3,100-foot level on veins exposed in old workings, and the 3,350-foot level was extended, but they could not extract the rock with any degree of efficiency because the means of doing so were too primitive and too ramshackle.

As the spring of 1982 began to soften the cold mountain air, the tide began to ebb. The 3,000 level was abandoned, as were the Carlton and Battle Mountain tunnels. Men quit and were not replaced. To John this could have but one meaning. The seal on the proceedings was set by an even more costly debacle: the Cresson decline.

The Ajax shaft, with all its macabre goings-on, was only a part of the production plan. The whole operation was supposed to mine 150 tons per day. Of this, 50 tons were to come from the Ajax, and 100 from the Cresson decline.

The Cresson was one of the famous old mines of the camp, now long defunct. In November 1914 the workings broke into a vug partly filled with an incredible treasury of gold. It may be that "the Cresson vug" was the richest single occurrence of gold ever found in North America. But that was long ago, and now only the rusting headframe and surface works stood near the head of Eclipse Gulch. The grayish-yellow rock dumps of the Maggie, the Gold Sovereign, the Trilby, and the Dante formed a backdrop on the hillsides behind. Above them rose the sweeping curves and pine-clad slopes of the hills forming the topographic center of the camp.

The owners of the Ajax had extended their holdings as the old mines closed, so that they now owned most of the camp, including the Cresson. The operating company drilled the ground around the Cresson, looking for ore near the surface. One hole intersected low gold values, from which scanty evidence they deduced the existence of an ore body. They decided to drive a decline into the hillside and mine this ore by blasthole stoping. They persisted in this latter decision, regardless of any indications that the rock might not be suited to this method of mining.

The decline portal was laid out so that it angled obliquely through a slope of rocky, earthy overburden instead of diving into the rock by the shortest possible route. As a result, it was difficult to start the portal in rock. A short distance underground, the new decline passed just beneath an old adit called the Texas tunnel and had to be timbered. At the time John appeared on the scene, the portal had just been completed, and the face of the decline was less than 100 feet from the surface.

With some difficulty, and much false economy, the company had cobbled together a drill jumbo from various odds and ends, mounting two big Ingersoll Rand D475 rockdrills. The D475s had gone out of production ten years before, and spare parts were as scarce as hens' teeth. With this device and a diesel scooptram, they set about driving the Cresson decline.

The rock was a yellowish brown, seamed with cracks, sticky, and with the consistency of concrete that is not quite hardened. Almost every round took its toll of drill steel and bits, which jammed irretrievably in the rock and had to be abandoned, to be bent like spaghetti by the next blast.

As the 12' x 10' decline pushed deeper into the hillside, the quality of the rock did not improve. John asked his superiors if they thought they would successfully blasthole-mine this rock. They said the rock would improve in the ore zone. He kept his mouth shut and attended to his duties.

Winter is hard enough in mining camps at the best of times. When mine workings are barely established and close to the surface, it is doubly so. On some days access to the Cresson was cut off by snow, but it soon evaporated or blew away and did not greatly hinder the work. For a period of about a month, however, scarcely a round came out of the decline.

The supply of water for drilling came through a buried line from the mill, half a mile away down the gulch, into a header tank, and thence underground. The drilling water line down the decline froze. As soon as they thawed one section with a propane torch, another froze. By the time they had the whole line running, it would be the end of the shift; in the interval between shifts everything froze solid. The cycle repeated itself day after day, and almost nothing was achieved until the cold became less bitter.

John remembered many a night, during that long, high-country winter, driving along the gravel road that led from the Ajax mine through a saddle between two hills and down to the Cresson. Warmly dressed in long underwear, a thick shirt, and quilted coveralls, he would park the company pickup beside the decline and walk over to the portal. Snow and frozen mud crunched underfoot; the air was burning cold on his face. The clear Colorado sky was ablaze with stars. The few lights around the portal emphasized the bulk of the cavernous ruined buildings of the old mine, dimly visible in the starlight.

The company paid no incentive bonus, and it was therefore sometimes a pious hope that the crew of three or four men would be at work underground instead of snugly ensconced in the heated shack that served as a lunchroom.

"Pard, we had to get that round drilled out, and we just now came up to eat."

"Pard, we just shot fifteen minutes ago."

"Pard, we wasn't quite sure what you wanted done about that ratty piece of ground, so we figured we'd wait until you came along to have a look at it. Matter of fact, we was expecting you an hour ago."

John knew that it was smoke-and-mirrors, and they knew that he knew, but the crew numbered fifteen or twenty men scattered through the Ajax mine as well as the Cresson decline. In the absence of an incentive bonus system, and with a crew that for the most part neither knew nor cared whether they would be working there in a year's time, there was not much that he could do. Provided that a certain reasonable amount of work got done, there was a mutual need to live and let live. The company had developed its policies on incentive bonus at a closely supervised, highly mechanized rock factory in another part of North America. It was about as different as anything could be from a crew of four men, with a jumbo and a scooptram, driving a decline under a starlit Colorado hillside.

The Cresson decline reached its target after advancing on a 12 percent grade roughly 1,000 feet from the portal in eight months. They turned and drove a level drift through the supposed location of the ore. Not only did the rock not improve, there was no ore. Nevertheless they drifted and raised and drove sublevels on the structures where the ore was supposed to be. The assays showed a bit of something that the drillhole had intersected, but that was all.

The rock was still sticky and continued to take its toll in "hung steel." John pointed out to Eric that, with that kind of drilling performance, longhole drilling was unlikely to be an economical proposition. Besides, he remarked, the ground was so weak that it might not support the open spaces that are inherent in that type of mining. If it caved, they would lose everything. John added that, even if he was entirely wrong, it was a risk that the company could ill afford at that stage.

Some mining methods (which do not figure in this tale) depend on the rock caving; others depend on it staying open. If the rock will not cave reliably when it is supposed to, or caves when it is not supposed to, the results can be a jackpot of alarming dimensions. In this case John did not believe that the rock would stand up in the required manner and suggested that the ore—such as it was—should be mined by cut-and-fill, using the rock dumps on the hillside as backfill.

The company had decided to blasthole-mine the Cresson and was not going to be swayed by advice from a mere foreman. They called in a consultant to advise them how to go about it; neither party gave any consideration to whether it was in fact possible. The company had a habit of doing that—spending $500 or $1,000 a day on a consultant to tell them what their own employees could tell them for nothing. If one of their engineers looked in a book and found that the cost of typical longhole drilling was so many cents a foot, they could not get their minds around the fact that sticky rock and jammed steel could actually cause that figure to multiply many times over. The cash-flow projections developed in eight years of studies conducted in the comfort of the office down the hill could not be sacrificed to the unpleasant realities underground. Theory and practice were set for a head-on collision.

The collision took place nearly a year after John had left the district, in the manner he had predicted. With great expenditure of money and effort the company managed to get the "ore body" drilled off. They blasted the first third of the rock they had drilled; the rest caved. No one was hurt. The crew went down there at the start of one shift and found that their handiwork had disintegrated into blocks the size of houses, making further work a waste of time. The company put a steel gate on the decline portal and walked away. The loss of the Cresson would play a significant part in the decision to abandon the operations at Cripple Creek. At the time of our tale, however, the shape of things was only beginning to emerge.

An intelligence of less than Einsteinian dimensions could become aware that the progress of the company's affairs at Cripple Creek was not smooth, rapid, or productive. The very week that John hired on, the company was taken over by another, larger, company. While the old company might persevere with what it had begun, the new corporate princelings would not be restrained by any such considerations. Several months passed before the first party of princelings appeared, looked around, and went away again. Whatever its scenic attractions, Cripple Creek was but a tiny element in the company's business. The axe would not be laid to the tree at once, but the time would surely come, and John and his fellows would, as it was elegantly put, be kicking horse turds down the hill.

John knew that one of three things would happen. The mine would kill him, the company would close the mine and lay them all off, or he would find himself gainful employment elsewhere. He set about the latter course as being preferable and put out the word along the grapevine of the mining fraternity.

His efforts finally bore fruit, and he found a job as the site engineer on a river diversion tunnel in the lotus-land of coastal British Columbia. It was only a temporary job, but a secure temporary job was preferable to working for an employer who swore on a stack of Bibles that no one would be laid off and continued to do so until the layoff slips came out with the paychecks.

One bright morning John met Eric in the mine yard. They were both avid students of mining history, and friendship had sprung up between them on their first acquaintance. In a way John had no wish to leave Eric to struggle with this unruly operation. The company was a first-class employer and an abnormally generous one. John knew, however, that events would overpower them, and the devil would take the hindermost.

"Eric, I hate to have to tell you this," said John, "but around here she's deep enough." The expression signifies the tramp miner's intention to pack his gear and move on. It is a statement of utter finality, beyond question or dispute, with no reason given and none asked. And so it was accepted.

Eric allowed his curiosity to get the better of him so far as to ask, with a sly grin, "What have you got in mind?"

"Tunnel job in British Columbia."

Eric paused for thought. He had worked on construction jobs before and had never been content to see only waste rock coming out of the ground with no hope of finding ore.

"John, how are you going to be happy with never any ore in the face?"

"There's never any ore in the face in this place, so what's the difference?"

Eric grinned from ear to ear. But it was true; there seldom was any ore in sight, and such a state of affairs could not go on for long.

Down at the office soon afterward, John passed Bertelsmann, who scowled at him but said nothing. Bertelsmann was used to cracking the whip over employees who were scared for their jobs and who revered the corporate hierarchy. In particular, John held it against him that he put Eric through the wringer at least once a week, causing the mine superintendent to tear about the place with a worried look on his face, talking to himself. John had had irate exchanges with Bertelsmann from time to time but had always fought fire with fire, never troubling his mind as to whether Bertelsmann enjoyed the treatment or not. It was perhaps a novel experience for Bertelsmann to have one of his staff throw his threats back in his face, and John's insolent grin did nothing to assuage his feelings.

John's last shift came one May night when the first warmth of the late mountain spring was in the air. The crew had gone home by a quarter to four in the morning with a flurry of running feet, lunchpails clashing against door jambs, shouts, revving engines, and tires scorching the gravel. John's job included locking the gate after the last man had gone.

In the quiet darkness he walked all around the surface buildings one last time and drove down to the gate. As he locked it, he looked back at the headframe silhouetted against the night sky. The mountain had tried to kill him, but he had escaped; he thanked God for his preservation.

Beside the gate stood the timber headframe of the Portland no. 1 shaft. John thought of Eddie Tape, now dead for some years, thanked his memory, and wished rest on his soul. He leaned on the roof of his car, resting his chin on his folded arms, and looked westward to the Sangre de Cristo Mountains, where the first thunderstorms of the year were in full swing. The clouds blazed from within with golden light as skeins of lightning sprang from one to another. All along the range lightning flared and shimmered fifty miles away in the quiet night.

21

IN THE HILLS THERE ARE GOLD MINERS

"In the hills there are gold miners," remarked the commissioner, tapping his front teeth with a pencil as he stood looking at a wall map that showed where the tunnel was. Delaney was a civil engineer in charge of a water utility in British Columbia. He was being swindled by the tunneling contractor and, on the grounds that it takes a thief to catch a thief, was about to employ John to put a brake on the swindling process. The subject on which he had just remarked was an unwanted complication.

Delaney was first and foremost an engineer, and a good one. In pursuit of his living he had become a bureaucrat, and an efficient one. His present capacity had shoehorned him into being a part-time politician, a role that he hated and that sometimes prevented him from sleeping properly at night. As an engineer and a bureaucrat, his world was clean, neat, and orderly; problems were solved by calculation, optimization, and careful design. It was a world where things happened because they were authorized to do so, approved by an appropriately qualified person. If Delaney and his kind had been entrusted with the creation of the world, they would have spent seven days selecting a short list of names for a steering committee to approve funding for a preliminary prefeasibility study of the project.

Delaney never came to terms with the fact that things happened because of the driving force of human will and ingenuity. The mere act of authorizing something never of itself caused anything to happen. People who made things happen, people who sweated and swore, screamed abuse at each other, bet their lives

on their own skill and cunning day by day, and then drank themselves paralytic or duked it out with the cross-shift on the sidewalk outside the bar, were not a part of this world—not authorized, not approved. Delaney knew that such people existed from having worked around heavy construction jobs, but he regarded them with detachment and distaste, although not without a certain awe. Gold miners were authorized—he could do nothing about that—but not approved. For this reason mystification and a quiet irritation were blended into the simple statement "In the hills there are gold miners."

Gold is not a rare metal in the West; in fact, it is common all through the western mountains from Mexico all the way up through Nevada, Arizona, California, Colorado, Montana, Idaho, British Columbia, the Yukon, and Alaska. What is rare is enough of it in one place to make it worthwhile spending $20 million on machinery and another $20 million on digging holes in the ground just to get at it. That is rare.

Half the creeks in British Columbia have been panned for gold successfully (and the other half unsuccessfully) at one time or another. The river running past the water utility's catchment area was no exception. Therein lay the problem. The river ran past the catchment, not through it. As the local population grew, it needed more water. The water utility needed to tap the river and divert part of its flow into the catchment area. The tunnel was to be the means of doing this.

The river ran in a forested canyon. In the darkness under the trees the ground was carpeted with fallen needles and scattered with dry boughs that would crack loudly underfoot. Where sunlight broke through, salal would grow in a knee- or waist-deep tangle, lit a brilliant yellow-green by the sun. The riverbed was anything from 40 to 100 feet wide. Boulders and slabs lay on ledges of bedrock; between them were shoals and pockets of sand and gravel. In summer the river could be crossed dry-shod. In winter it became a foaming, bank-full torrent raking its bed.

In the 1860s prospectors found gold in the riverbed. A brief flurry of excitement followed, but the river offered slim pickings compared with other placer diggings, and the boomers moved on almost as soon as they had arrived. Even the Chinese did not stay for long. Nevertheless a few placer miners—men and women—lived from decade to decade in cabins scattered through the woods and did the best they could.

They were quiet, private people and as shy as the deer. Yet to those who tiptoed up and knocked on the doors of their lives, they were warm-hearted, generous, and hospitable. After many meetings by chance or by design, and after the visitor had been offered many cups of coffee and slices of home-baked pie, a pill bottle would come out from under a mattress or from behind a cupboard, and in the pill bottle there would be river gold.

They paid no property taxes, mortgages, or utility bills, had no bank accounts, borrowed no money, had no washing machines or televisions. Their music was the wind in the trees and the rushing of the river. Their entertainment was the ever-changing light and shadow of the valley. Some of them hiked down to the village to backpack their supplies into the woods. Some of them clattered and lurched down the logging roads in ancient pickups for that same purpose. The logging companies paid them a little every summer to cruise the woods, watching for fires. They worked their claims with pans and sluices, using crowbars to shift boulders so that they could burrow for pockets of gold underneath. From time to time they would scrounge a stick of dynamite, a blasting cap, and a fuse to dispose of a big boulder beneath which gold nuggets might lie.

Almost all placer camps have traced their rivers and creeks up to some deposit in solid rock, and hardrock mining has often followed. Here, where John was to take up his new employment, many had looked for gold in the bedrock and had gone on doing so for a hundred years, but without success. No one had come up with a convincing idea of where the river gold might have come from. The theories put forward lacked either proof or conviction.

Not all the gold miners were quiet, gentle folk. One of them was a man named Jameson, who was a true brigand from olden times. In another age he would have been a pirate, a highwayman, a rumrunner, or a gunslinger. In another age, too, or south of the border in this one, his ventures and adventures might have been ended by a lead lunch. But this was Canada—peaceful, law-abiding Canada—where even the possibility of such a solution was considered fit only for those who were incapable of solving their problems by due process of law. Nevertheless, by due process of law, the man flourished for a while and made a goodly dollar for lawyers, geologic consultants, and, indirectly, for John. The story is long, drawn-out, and devious, so we will skip the dull bits and follow the inscrutable workings of fate.

The story began—or rather, a series of events began to unfold with enough coherence to be called a story—nearly ten years before John arrived on the scene. Even though no one had ever found gold in the bedrock in a hundred years, Jameson restaked some old claims on the bedrock—not placer claims, but lode claims—and then staked several new ones. At that time the water utility had it in mind to tap the river by laying a pipeline (the tunnel idea came later), and perhaps it was coincidence that the claims covered a part of the pipeline route.

Some of the roads to Jameson's claims ran through the utility's catchment area. The water catchment area was closed to public access for sanitary reasons by means of locked gates. The gravel roads within it were narrow and sinuous. The utility's vehicles were fitted with two-way radios, by which the drivers broadcast their position and intentions. It can well be imagined that, if everyone drove about on the assumption that they knew the whereabouts of oncoming traffic, a vehicle without a radio, driving within the water catchment unescorted and unannounced, was a hazard to all concerned. Jameson gave scant attention to the safety of others, locked gates, regulations, or any other impediment to his activities. He broke through gates and moved his equipment through the catchment area whenever he pleased. The utility uttered threats but was unable to take any effective action. Jameson continued with impunity.

Unlike a mine, a water utility is about as permanent and indestructible an organism as any that can be imagined. If anything survives the end of the world, it will probably be a water utility. The need to tap the river had been foreseen for twenty-five years; growing pressure of population brought this need progressively out of the distant future and into the present. The decision was taken to tap the river by means of a ten-foot diameter tunnel rather than by a pipeline.

One reason for this decision was that the commissioner who was Delaney's predecessor had been a mining engineer. He had worked for Cominco at the Box mine, on the shores of Lake Athabaska, in the years before World War II. Returning from wartime military service, he married a girl who very sensibly refused to live on the shores of Lake Athabaska, or anywhere similar, so the mining engineer earned a degree in civil engineering. He became the first commissioner of the newly formed water utility, and they lived happily ever after—they really did. But once a miner always a miner, and the commissioner had a propensity for driving tunnels.

As a result the charming and civilized employees of the water utility were obliged to invite into their peaceful midst the rude, crude, and generally undesirable characters who apply their violent energies and ruthless cunning to the business of blasting holes out of the rock.

Once the water utility had planned its tunnel route, it was required by law to inform all nearby mineral claim holders of what it intended to do, and where and how it intended to do it. Jameson, as a claim owner, therefore received a plan of the proposed tunnel route by registered mail. At the same time, the utility applied to the provincial Ministry of Mines for a mineral reserve, which would prevent claims from being staked over the tunnel route. Six months passed before the request was granted. In that time the utility surveyed the tunnel right-of-way and deposited the drawings in the Land Title Office as required by law. Within days of the deposition of the drawings, Jameson staked claims all over the tunnel route, which the Ministry of Mines recorded and accepted.

It is difficult to find ground with enough mineral in it to be worth mining. Once the ground has been found, it is even more difficult to prove that its mineral contents can be extracted at a profit. In fact, it cannot be proved in advance of actual mining, only predicted on the basis of available information. The predictions are often wrong, to the acute embarrassment of all concerned. It is even more difficult, however, to prove that a piece of ground does *not* contain valuable minerals. In fact, it cannot be done. Jameson's lawyers were aggressive and astute; the water utility's lawyers had neither of those attributes. The utility squirmed and struggled. Geologic examinations were negative. The driving of a tunnel beneath this piece of moose pasture would enable its geology to be investigated with a degree of thoroughness seldom accorded to any similar piece of moose pasture ever before. Nevertheless, Jameson demanded a million dollars for the trespass that the tunnel would commit on his mineral claims.

The affair dragged on for more than three years. In the end the utility decided to expropriate the claims, and an arbitration committee was set up to determine the compensation that the water utility should pay to Jameson. The committee consisted of a chairman and one representative from each side. The water utility's representative was delayed on the day of the hearing. He arrived to find that the chairman and Jameson's representative had reached an agreement. He signed on the dotted line, went home, and invoiced the water utility for his valuable services.

The arbitration committee awarded Jameson a sum in five figures and required the utility to engage a consultant to map every foot of the tunnel, take samples for assaying where appropriate, and provide the results to Jameson. The committee also laid down what was to happen if valuable minerals were to be found while driving the tunnel. Thus Jameson was paid for investigating his own claims at the water utility's expense.

The tunnel went ahead, which had, of itself, nothing to do with John, who was at Cripple Creek, slugging it out with the gas, ghosts, and things that went bump in the dark. But the hand of fate is often busily at work setting the stage long before the actors have been given their lines.

When John hired on at Cripple Creek, a Philippino mining engineer by the name of Arnold Cabrera was employed there as a safety officer and in various other capacities. Because Arnold and John were both outsiders, both foreigners, they shared thoughts and information that they shared with no one else. Arnold confided to John that he was not happy in the mountain fastnesses of Cripple Creek and intended to return to his family in Las Vegas. John confided to Arnold that he could see little future in the operations at Cripple Creek and intended to return to Canada. Arnold departed and was seen no more in those parts.

The consulting firm that made some of the geologic investigations for the water utility saw that the tunneling contractor was leading their client around on a string. The old commissioner had retired, and the new commissioner, Delaney, was also aware of this problem. At that time the consulting geologists needed tunnel inspectors for the construction of a railroad tunnel in the mountains and placed an advertisement in the *Vancouver Sun*. Arnold Cabrera had a Philippino friend in Las Vegas who, for reasons best known to himself, subscribed to that newspaper. He saw the advertisement and passed it to Arnold. Arnold mailed it to John. John had no idea what a tunnel inspector was—he thought it was something to do with sewers—but applied for the job anyway, reasoning that it is impossible to throw a six without first rolling the dice. By that time the positions on the railroad tunnel had been filled. The consulting firm passed John's application to their client, Delaney.

On receipt of this information from the consulting firm, John telephoned Delaney. Delaney pointed out that the contractor, who had started the tunnel by drilling and blasting, was about to install a fabulous mining machine that would complete the tunnel in eight months. He added that the position was a temporary one, that it was hardly worthwhile for John to move from Colorado for an eight-month job, and that an older man, in semiretirement, was about to accept the position anyway. John replaced the receiver on the hook with a philosophical sigh.

A few days later the telephone rang. It was Delaney. The man who had been about to take the job had refused it because of its probable short duration and because it would require him to move house. Was John interested? In the course of the conversation Delaney had described the rock in which the tunnel had been driven to date and its probable continuation. John knew that fabulous mining machines of this particular variety worked well only in certain situations; a hardrock tunnel was unlikely to be one of them. He guessed intuitively, even as he and Delaney talked on the telephone, that the machine would not work as claimed, if at all, and replied that he would take the job for as long as it lasted. John took a couple of days' vacation from Cripple Creek and made a quick trip to British Columbia. Delaney showed him the site and offered him the job, which he accepted. John returned to Colorado, called it deep enough, and hit the road for British Columbia.

John arrived on the site at the same time as the fabulous mining machine. The machine was a dismal failure, driving the tunnel a mere six feet before the contractor reverted to drilling and blasting. Driving of the tunnel was completed two years later. The utility decided to line it with cast concrete. Rather than engage a contractor, they decided to do the job themselves, and John found himself in charge of a $4-million job to line the two-mile tunnel with 17,000 cubic yards of cast concrete. So John's chance eight-month job lasted for four happy years, but that is another story.

The driving of the tunnel went on. Jameson never got his million dollars. One law firm after another, working for a contingency fee that faded ever wraithlike before them, slid him out through their doors. The water utility's lawyers probably raised their families on the proceeds. Because each foot of the tunnel had to be mapped, the consulting geologists were happy. The water utility was happy because it was no longer being swindled. The contractor was not happy. Jameson brooded and bided his time.

John was very happy. The mining industry was sinking into a bitter depression from which some said it was unlikely to recover. Many mining people considered themselves lucky to have jobs at all in places like Yellowknife or Thompson. Some found themselves out of work and living on beans and bacon-grease sandwiches. It was like musical chairs; the music stopped, and it was hard times for those without a chair.

News came over the grapevine from Cripple Creek that half the crew had been laid off three months after John's departure. Less than a year later, everyone, including Eric and Bertelsmann, went down the hill kicking horse turds.

John, meanwhile, had a cushy and well-paid job in a place that closely resembled his idea of paradise. He reckoned that his experiences at Cripple Creek had left him with his dues paid up for years to come, and he thoroughly enjoyed life. He remarked to a friend that he was tempted to marry a Russian princess and retire to an enchanted island, both of these being available in modest numbers. John never met the right Russian princess, nor did he find his enchanted island. Perhaps it was just as well, for the tramp miner is an incorrigible breed, hopelessly addicted to the ultimate narcotic—adrenalin—and is forever pushing on in search of the next impossible challenge to provide his fix.

For two years all was quiet. Jameson took on an almost mythical quality as some wild creature of the woods, and John never met him or wished to. But then came Friday the Thirteenth.

The tunnel had been driven to a point twenty-five feet short of breaking out on the riverbank, where a ledge was blasted out as a portal site, and that was how it was to remain for the time being. Fifteen feet of rock separated the roof of the tunnel from the ledge on the riverbank, which could be reached by a rough road zigzagging down the side of the canyon. The contractor's men were at work in the tunnel, about 400 feet from its end, grouting off some water-bearing fissures.

Grouting was a cold, wet business and deadly dull. John's job as site engineer entailed nothing more than standing there for the duration of each shift, recording holes drilled and bags of cement pumped into fissures in the rock, with almost the same intensity of boredom and discomfort as the vigil of the drill drift ten years before.

It was early afternoon, Friday the Thirteenth.

The crew heard a dull crash from the supposedly empty blind end of the tunnel, half-heard over the ululation of the grout pump.

"Hey, what was that? Turn that thing off!"

They listened. They heard another crash, as of falling rock. John and one of the miners walked toward the end of the tunnel to investigate. They had only gone halfway when the sound was repeated more loudly, followed by an intense sibilant roar as of some fluid entering the tunnel under high pressure. The two men stared at each other, aghast. Their one thought was an inrush of water. There was no reason why there should be any such inrush, but it was two miles to the portal, and if a man thinks that he is about to be drowned, he does not waste time asking questions of a technical nature, or indeed of any nature.

In terror they ran back to the grouting crew, yelling that they were all to vacate the premises without delay. In terror they sprang onto the locomotive, abandoning everything. And in terror John looked behind him for the foaming flood that was about to overtake them as the locomotive ran at full speed toward the star of light that was the portal.

They shot out into the daylight. The locomotive skidded to a halt. They sprang off and scattered, looking back at the portal. They stopped. Everything stopped. Time stopped. Nothing happened. Nothing came out of the portal, not a sound. Not a trickle of water was added to the steady flow in the ditch. The mechanic wandered over, grinning and wiping his hands on a rag. "What's with you guys?" he asked. In the quiet of that warm spring afternoon, out there in the sunlight, they were not sure of the answer.

John and some others drove around to the canyon where the tunnel would eventually emerge. Looking down into the gorge, they could see a cloud of dust hanging on the still air and could hear the clatter of a big quarry drill. Someone had set up an Airtrak and was drilling into the tunnel.

John walked down the road toward the drill with Herk just behind him. Herk worked for the water utility and had the happy ability always to be in the right place at the right time. He was an unobtrusive individual, but if you looked at him a second time, you would realize that he stood somewhat over six feet in height and weighed something like 220 pounds without an ounce of fat on him. Herk was a good sort to have around at such a time.

John had a feeling of unreality. This kind of thing did not belong in Canada in the 1980s. This event would have been appropriate to any number of places—Deadwood, South Dakota; Creede, Colorado; Tombstone, Arizona—and any time between 1860 and 1920. It was just not fair that it was happening here, now. Such things were quite definitely not authorized, not approved.

With Herk's reassuring presence behind him, John approached the driller and his partner.

"Are you Jack Jameson?" he asked the elder of the two.

"Yes I am," the man replied, eyeing John balefully.

"What are you doing?" asked John.

"Drilling some holes."

"Why are you drilling those holes?"

"That's my business."

John looked the setup over. The way the holes were being drilled, it would be a sure thing that, if loaded and blasted, they would blow the end of the tunnel wide open and possibly let in the river. No one could work in the tunnel under that threat. Herk and John could do nothing further, so they went back the way they had come.

That Friday evening the water utility was like an overturned ants' nest. They contacted the police, who told Jameson to cease and desist. Jameson called it quits for that evening and went to talk to his lawyer. He neither ceased nor desisted but went on drilling the next morning. In response to the utility's complaint the police said, pusillanimously, that it was "a civil matter" and washed their hands of it. It being a Saturday, lawyers and justices of the peace could not be found. Everyone made themselves scarce, leaving John and Delaney holding the bag.

The two of them drove up to the site, that warm Saturday afternoon, along a road that ran through the trees across the river from where Jameson had been drilling. All was quiet. Delaney sat in the car while John crept down through the trees to the riverbank. If the blast was fired, the trees would provide a degree of shelter from flying rock. The site was deserted. In John's experience deserted places were of two kinds: places that were just plain deserted, and places that were deserted because someone had gone away to fire a blast. Dust coated the ledge beside the river, where two or three dozen red plastic plugs marked the holes. There was no sign of life, nothing stirring, utter quiet except for the carefree rushing of the river. John sniffed for the smell of burning fuse and looked for telltale plumes of smoke. He saw none. He looked for primer leads, which otherwise must surely be there if the holes had been loaded for a blast. He saw none. Jameson and his equipment had vanished.

Delaney and John drove to the tunnel portal, where two pails of quick-setting cement were kept, intended for use in the grouting work. They collected these, a bucket, a length of drill steel, and some rags and newspaper. They drove directly to the site, meeting no one. They did not have nearly enough cement to fill all the holes, so they mixed what they had, rammed wads of newspaper two or three feet down each hole, and filled the tops with cement. The next day Delaney had some of his employees dump several truckloads of earth and rocks all over the site. Jameson's guns were spiked. Even if he cleared the rubble away, he would still have to drill out the holes before he could load them. The sound of any such drilling would carry through the rock to anyone listening inside the tunnel.

Each morning John went down to the riverbank but found no one. He sipped a cup of tea from his thermos and let the peace of the valley in the morning light sink into his mind. He returned to the portal, went underground with the grouting crew, and spent half an hour at a time sitting in the end of the tunnel, listening for the sound of drilling. Nothing happened; work in the tunnel was finished for the time being, and it was as if the whole affair had never been.

It would make a better story if the affair had ended with some clash of titans, but it was not so.

Utilities, almost by definition, deal with things that run in lines—railroads, power lines, pipelines. These lines occupy legally titled strips of land known as rights-of-way. The land covering the tunnel route was divided into four or five different, but contiguous, rights-of-way butted end to end. The arbitration agreement between Jameson and the water utility covered a specific area of land, no more and no less, defined in terms of certain specified rights-of-way.

The last twenty feet of the tunnel extended across the boundary of one right-of-way into another that covered the site intended for the intake works. Through the incompetence of the water utility's lawyers, this particular right-of-way had not been included in the arbitration agreement. The water utility therefore could not carry out work on the upstream end of the tunnel without entering into further negotiations with the odious Jameson, who had probably known of this defect in the agreement all along.

The fact that the tunnel had been driven for 11,000 feet without encountering a trace of valuable mineral of any kind cut no ice in the courts and did not constitute proof that this particular right-of-way, even though it measured only some 200′ x 200′ and contained abundant outcrops of barren rock with no sign of mineralization, did not contain valuable minerals. The consulting geologists' final report, which said, in technical language, that there was not a snowball's chance in hell of any valuable minerals anywhere in the vicinity, for some reason never appeared as evidence. "And whereas my client . . . etc., etc."—and so it went on for another eight months. The utility paid Jameson off with another sum in five figures and spent more money on legal fees, and the whole affair ended with as little fanfare as it had begun.

The story had one little quirk. In the contractor's site office was a calendar. The Friday after Friday the Thirteenth was Good Friday. No one would have been at the tunnel on that day or for three days following. The contractor's site superintendent had marked that day with a red circle. But Jameson had jumped the gun.

22

THE MINE AT BURNT ISLAND POND

All tunneling jobs come to an end, and John's was no exception. The spring of 1987 found him hanging out his shingle as a mining consultant and sharing an office in downtown Vancouver with two friends who had been partners with the water utility's consulting geologists and now had their own business as consulting geologic engineers. Just as John found his way into heavy construction by accident, so now he found himself in the thick of another industry of which he had hitherto been only dimly aware—mineral exploration.

John had spent his time in mines that had been discovered anywhere from 50 to 1,000 years ago. He had never bothered his mind with the details of how mines are found today. His idea of mineral exploration had something to do with geologists banging on rocks. It had never occurred to him that mineral exploration is a substantial, if obscure, industry in its own right.

It is a strange, quirky sort of industry. It has no "product" in the manner of a mine, a smelter, or a factory. Its product is information. This information is acquired at great cost in the most arduous conditions out in the wilderness, and most of it is negative. The industry is fueled by money from the exploration budgets of large mining companies and from people who buy stock in "junior resource companies" in the hopes of making a profit when they sell the stock. Most of the money invested in mineral exploration is lost. A tiny fraction is multiplied hundreds or thousands of times over when a mine is found and brought into profitable production. The exploration divisions of

big companies work in secrecy. Small companies trumpet their successes in the hope of raising the price of their stock, rewarding their investors, and generating money to keep looking for the big one.

North America is overcrowded only in certain limited areas. We need only fly from Denver to San Francisco, Seattle, or Europe in clear weather to realize that much of the continent is no more settled than it was a hundred years ago. This wilderness, especially in Canada, is where the exploration industry functions. In the United States, exploration is more likely to cover preexisting mining areas because the country has a longer history of mining and has been more thoroughly explored in the past.

Let us suppose that we have found a piece of moose pasture where we think there may be ore. We cannot go into the reasons for thinking this; suffice it to say that we do. We stake claims or take an option on existing claims. Bear in mind that this is likely to be primeval wilderness—nothing but forest, rock, and swamp. Work crews may have to be flown in by helicopter or floatplane and must somehow be fed and sheltered. Some kind of trail may run through the area, and a cushy job may even be near a small town with a motel.

"Staking claims" means exactly that—slogging through the squelchy wet bush, clambering over deadfalls, falling into bog-holes, trying to keep moving in a straight line, trying to keep track of where you are, and marking the claim corners. Each day you do not believe that the bush can get worse; each day you are wrong. If the weather is fine, you are lucky; if it rains, you are going to get even wetter than usual. The insects are plentiful and vicious. You end each day ravenously hungry, wet, filthy, and exhausted. When staking is complete, the claims are recorded at the government mining recorder's office.

A geologist fights his way through this same piece of bush looking for "outcrop," where bedrock is exposed. In some places there is nothing but outcrop, nothing but rock, ice, and scree. In other places there is no outcrop, only lakes, muskegs, and slow-flowing creeks. Whatever the geologist finds, he records it.

The geologist may recommend geophysical or geochemical surveys. This means that grid lines must be cut and marked across the claims for the surveys to follow. It is no use taking hundreds of measurements if it cannot later be established where each one was taken. A geophysical survey consists of humping various electronic devices across our piece of moose pasture, stopping, taking a reading, recording the result and the time and place where it was taken, and moving on down the grid line, which is marked with pieces of bright plastic ribbon tied to trees and shrubs. A geochemical survey entails a similar procedure, except that hundreds of small samples of soil are taken and wrapped in labeled plastic bags.

The results of this work will be hundreds, more likely thousands, of numbers plotted on a map, which may give some indication of an anomaly—an area of abnormal readings—which may be worth pursuing further. The interpretation is as much an art based on long experience as an exact science. Can't you just stuff it all into a computer? Not if you value your professional reputation. A computer can ease some of the error-prone, brain-busting toil of plotting results. By being able to juggle and combine numbers very quickly, it may reveal patterns that would otherwise remain hidden, but that is all.

All this work—and we haven't got very far, have we?—is probably concentrated on a piece of ground a mile or so across. It will have cost something like $50,000, even $100,000 for a big program. The answer may come back, gift-wrapped in geologic technicality, that the claim is still a piece of moose pasture. The tough properties are the "teasers," showing a little mineral here and there. They can consume far more than $100,000 and still be moose pasture at the end of it.

In Canada much exploration work is restricted to the summer months by snow cover and weather conditions, and there is talk of "the exploration season." If you earn your living in mineral exploration and this year's season is quiet, you will be frying your bootlaces before next spring—assuming that you can afford the grease to fry them in.

Work on a "property" goes ahead season by season at a rate determined by what each season's work turns up and by the ease with which money can be raised. If the price of the metal being sought is high, or if work turns up good results on this property or another nearby, raising money will be easy. Therein lies one of the absurdities of this business. If, for example, the price of gold is high this month, investors will flock to a good gold showing. Few properties, if any, can be developed from raw wilderness to a producing mine in less than five years. By that time the price of the metal will have been through many fluctuations, so today's price is almost irrelevant, but that is how it seems to work.

Governments, too, have their heavy-fisted, fumble-footed influence, which tends to be cyclic. The cycle consists, first, of attempts to kill the goose that lays the golden eggs, followed by a realization of the enormity and foolishness of that intention. We could continue in a cynical vein, but let us refrain with the comment that this cycle seems to have a period of about ten years.

Each successive work program on a property, year by year, produces a mass of information that must be accurately recorded and compiled. The interpretation will result in a recommendation as to what further work should be done to develop our understanding of the property. If no further work is recommended, the option will be dropped or the claims abandoned—or sold if possible. A property may lie dormant for years or decades until someone comes along with a new interpretation of the geology, or in search of a different mineral. But let us suppose that the results of exploration continue to be favorable.

The next stage consists of diamond drilling, although, at the same time, more claim staking and geophysical and geochemical work may go on in the surrounding country. It is said that one raw prospect in a thousand reaches the diamond-drilling stage. Of a thousand diamond-drill prospects, one in a thousand makes a mine. Not every mine makes a profit, so the business of making a profitable mine is a tough road against long odds.

Our property, a few years from when we first looked at it, is still raw wilderness, but now we are going to move several tons of machinery onto it and live there for a month or more. Everything needed has to be trucked or flown in. Diamond drilling is done by specialized contractors. A program may cost something like a quarter of a million dollars, sometimes more, depending on the number and depth of holes that we require. The result will be a stack of wooden boxes containing drill core—and more information. That is all.

The diamond driller's life on the site is centered on a machine with rapidly rotating parts, plastered with black grease, emitting a steady, deafening whine, and powered by a blaring diesel engine. Diamond drillers used to be identifiable by their missing fingers. The rig may be surrounded by a shack of tarpaulins or plywood, providing some degree of shelter from wind, rain, and snow. The monotonous routine of adding drill rods to the string as it bores steadily into the rock and then pulling each ten-foot run of core is broken only by pulling the rods on completion of the hole and moving to the next location, perhaps a few hundred feet distant. All of this is recorded on a drill log, a document of lines and columns with sparse entries in rude calligraphy, covered with black smudges and greasy thumbprints.

The driller's abode is a tent that he shares with his partner. A few planks form the floor; a crate serves as a bedside table. The furniture is completed by two camp beds, each bearing a smelly sleeping bag. A clock, a radio, a week-old newspaper, a worn pack of cards, and a bag bursting with dirty clothes complete the driller's possessions. The camp may include a cook tent staffed by that legendary eccentric, the camp cook, but on a small job the crews are thrown back on their own culinary skills. Such skills blossom in the most unlikely people. Washing facilities tend to be somewhat basic, but healthy people do not stink when they work hard in the fresh air; out in the bush everyone smells the same anyway.

Drilling goes on around the clock, seven days a week. On a short job the crews stay on the site for the duration. On a longer one they may be rotated in and out. After one to two months of a hard, unvarying routine, constant dirt and discomfort, and the same faces to look at, men become "bushed." The first symptom is the loss of a sense of humor. Some people have no sense of humor and do not belong in a camp. But even the man who is the life and soul of the crew when fresh from "outside" eventually becomes dull, morose, and irritable. In extreme cases some innocent prank or minor frustration can provoke outbursts of violent rage. More usually, the other inhabitants of the camp treat the bushed individual with care and look forward to his turn to go out even more eagerly than he does. There is no room for glum faces around the camp.

Who are these people anyway? They are contractors and their employees. Whatever the size of the company doing the exploration work, it is unlikely to employ people with all the necessary skills or to possess all the necessary equipment. These deficiencies are made up by engaging contractors—line-cutting contractors, geophysics contractors, diamond-drilling contractors, and perhaps, eventually, mining contractors. The client

wants to be able to call up experienced and competent personnel, properly organized and equipped, get the job done, and
then pay them off. The contractor is, by definition, content with
an erratic way of life, always moving from job to job, finding his
own sustenance and creature comforts. Contracting offers the
possibility of making a fortune or going bankrupt; it is an arena
of merciless competition in which the weakest go to the wall.
The demands and corresponding satisfactions are infinitely
greater than anything offered by the settled routines to which
most people are accustomed. As a result it attracts the free spirits—tough, restless, self-confident men, adventurers, and, by
definition, tramp miners.

All this hard work and all this money produces information,
nothing more. The information must be plotted on maps, plans,
and cross-sections with all previously acquired knowledge.
Gradually the grade and tonnage of the mineral deposit becomes clearer as diamond drilling adds a third dimension—
depth. Typically the summer months are spent gathering information, which is plotted and analyzed during the winter. In the
early months of the new year the next summer program is designed and budgeted, and funding is obtained. The yearly cycle
repeats itself. At the diamond-drilling stage the question is
heard for the first time: "Can this deposit be mined at a profit?"
Even with a promising deposit, several years and more than a
million dollars may be spent just to reach this question.

With some ore deposits a valid decision to make a mine or
not can be taken on the basis of diamond-drill information
alone, but they are rare. The next stage, normally, is to go underground and inspect the ore in its bulk state. This may still be
necessary even if the final result is an open-pit mine. If the deposit is in a hillside, an adit can be driven, measuring, perhaps,
8' × 8' up to 12' × 14' in size. In flat country a shaft must be
sunk, or more often an inclined tunnel, called a ramp or decline, is driven downward on a 15–20 percent grade for a slope
distance of 1,000 or 2,000 feet or more.

This is no small operation. The job will not cost less than a
million dollars. The minimum crew will be fifteen or twenty
men. A hundred tons of equipment may have to be moved onto
the site. Resupply will probably be possible in the course of
the job, but everything needed to get the work started must be
moved onto the site at mobilization—not just machinery but

such things as electrical fittings, cable, welding rods, nuts, bolts, explosives, diesel fuel, lubricants, not to mention the camp, food, bedding, soap and toilet paper, basic office equipment and supplies, and the means of providing heat, light, cooking facilities, sanitation, drinking water, and telecommunications.

The work of assembling this equipment and moving it onto the site is a logistic operation of formidable dimensions. The amount of detailed information that a contractor's personnel must have at their fingertips or stored in their heads is enormous. How many main assemblies does an ST-5A break down into, and what is the weight of each? What are trucking rates between Sudbury, Ontario, and La Ronge, Saskatchewan? Who is the Atlas Copco representative in Yellowknife? How much and what kinds of food do twenty men need over five months? What are the provincial regulations concerning site effluent? What are the lead times on the necessary permits? Questions like these must be answered in their thousands in a very short space of time. The one thing worse than not knowing the answer is forgetting to ask the question. The answers go into a contractor's bid price for doing the job. The bid price is a major determinant of which contractor gets the job, as the company doing the exploration will most likely invite tenders from several contractors.

The job is always under time pressure of the most extreme kind. Quite apart from the fact that the contractor's profit depends on doing the greatest possible amount of work for the least possible cost in the shortest possible time, work must sometimes be completed in summer before winter weather clamps down, or in some parts of Canada it must be completed in winter before ice roads thaw and land access becomes impossible. Delays of the most infuriating kind interfere with mobilization— permits stuck on the desk of some government official who has gone on vacation, or a client who cannot raise the money or cannot make up his mind what he wants done. Time and seasons march on, and the weather may break early this year.

In the end the contractor's site superintendent is left holding the bag. It is his responsibility to get the job done on time, within budget, without breaking too many regulations too obviously, to keep his men happy, to keep the client happy, to keep government inspectors happy, and to keep himself from going insane. The physical and mental stress of this position is enormous. (So is the pay that goes with it.) Contract mine development out in the bush is the tough end of a tough racket. Fools, wimps, and dithering incompetents need not apply.

But let us return to the partnership in Vancouver. It was the spring of 1987; mineral exploration by junior companies was booming all over Canada, encouraged by successful government tax incentives, a good price for gold, and high prices on the stock markets. New mines were continually being found, and everyone in the industry was hard at work making money and having a good time.

John's two geologist friends, Robert and Clarence, were partners in their own company, which they had named Canroc Mineral Consultants Ltd. John was their associate. A friend and retainer of long standing named Jim acted as bookkeeper, expediter, and general factotum. They all shared a comfortable office with their names on the door and could look forward to the 1987 exploration season with optimism.

One of their clients had a gold prospect in Newfoundland, an island off the Atlantic coast of Canada, 3,000 miles from Vancouver. The property had a history of exploration dating back ten or fifteen years. One company after another had worked on it for a year or two and then passed it on. The property definitely contained gold, and the property map was patched like a quilt with the results of geochemistry, geophysics, and diamond-drilling programs. Two or three summers before, a contractor had driven a short decline and had drifted on an ore-bearing structure before the property was once again abandoned.

Canroc's client had subsequently acquired the property and had done some exploration work in the summer of 1986, the results of which were being plotted on maps in the company's Vancouver office. The plan for 1987, which Canroc helped to design, was to pump the water out of the old decline, drive 1,000 feet of fresh decline to gain access to the mineralized zone 150 feet from the surface, and then drive 1,000 feet of drift along the zone so as to obtain a bulk sample of about 1,000 tons. At first the plan was that Canroc would engage a contractor on the client's behalf and oversee the work, but John and his two friends put their heads together and made a proposal to their client that, for suitable remuneration, Canroc would be able and willing to undertake the whole job. The client was attracted by this idea and, after due consideration, engaged Canroc to carry out the work. Clarence would act as general manager in Vancouver; John was to be the site superintendent.

Canroc's office in Vancouver at once became a hive of activity as they laid out the project, finalized the budget, and wore out the telephones scouring Canada for rental equipment. A journey of reconnaissance was needed to view the site. In addition, John needed to inspect the equipment that they were to rent, which was located in various parts of eastern Canada.

The client company maintained a camp on the site, which was separated from the nearest road by ten miles of almost impassable bog and rock. Helicopters were the only available means of access. In this way Clarence and John found themselves in a helicopter, flying through the gray skies of Newfoundland on a cold, wet June afternoon.

The helicopter pilot picked his way between the clouds and fog banks that lay on the southwest coast of the island. A white dot on the moors came into view that resolved itself into a primitive camp as they approached it. The smallness of the camp emphasized the grandeur of the sweeping curves and slopes of the brown moorland. The helicopter landed beside a collection of fuel drums, generators, and assorted supplies near the camp. They climbed out onto the muddy ground. The subsiding whine of the helicopter's turbine gave way to a silence broken only by the wind. As John looked about him, the scene imprinted its profound desolation on his mind.

All about them spread the rolling moors, clothed in brown grasses and moss that squelched underfoot. Where the mat of vegetation had been torn by human endeavors, it revealed waterlogged black peat that could turn from mud to dust, and back to mud again, in a matter of days. In places the peat had been plowed and churned into a black morass, yet bedrock was close to the surface and outcropped in mounds and hummocks and low, rocky tors. The camp itself stood on a knob of bedrock, barely 100 feet across, which broke surface in the camp yard. What would appear, at a distance, to be a low green mound would turn out, on closer inspection, to be a clump of trees no more than six or eight feet tall, planed off by the force of the wind and matted so closely together for mutual protection that it was almost possible to walk on top of them. Meres with banks of spongy moss lay in hollows. One of them—John never found out which—was known as Burnt Island Pond.

The camp had been dragged in over the bog by the hapless contractor who had excavated the portal and 750 feet of 10′ × 10′ decline some years before and had lost money in doing so. It consisted of trailers, arranged on four sides of a square, surrounding a yard of stones and mud. A wooden boardwalk ran around the inside of the square. The contractor had abandoned the trailers on site; weather and casual vandalism had taken their toll. One trailer had been overturned by a winter storm. The aluminum siding of the others was torn and scraped, spattered with mud, and smeared with dust. Doors, windows, and interiors were in the last stages of dilapidation and decay.

Two or three men were busily sawing and hammering amid the squalid wreckage. A cook in a white apron was preparing their supper and dispensing hospitality to all and sundry at a gas range in one of the trailers, which was fitted out as a cookhouse. Its neighbor was the eating room with benches and two long tables, each bearing a cluster of jars and bottles containing pickles, sauces, and jams. John realized that he would be living in this place for several months and found the idea remarkably unattractive.

A road had been cut down through the peat to bedrock and led for 1,000 feet downhill to the portal of the decline. Beside the portal was a rock pile, some scrap metal strewn about as if the site had been blasted by a bomb, and that was all. Remnants of winter snow still lay in the open ramp leading down to the flooded portal. Rusty pipes ran at an angle into the scummy water. Below the black rock dump a slope thinly wooded with twisted trees fell away to the Isle aux Morts River, which could be heard rushing over its stony bed as its sound, echoing in the rocky valley, came in gusts on the wind.

John imagined miners coming and going in their yellow oilskins, hard hats, and rubber boots, scooptrams growling on the ramp, the siren howl of a ventilating fan, the blare of diesel generators and compressors, the crackle of someone welding, and all the hustle and bustle of an active construction site.

He reflected, with an alarm verging on terror, that the transformation of this peaceful but desolate scene into one of intense activity was his responsibility. Something would arise out of nothing because he, the magician, would bend his will upon it and make it so. The transformation had, moreover, to take place in an incredibly short length of time. Several large machines and quantities of assorted hardware would have to be brought from as much as a thousand miles away. If one piece of the jigsaw was missing, or was of the wrong size or shape, serious delays would result. He had to imagine each activity, each subsystem, and each line of supply, follow it in his mind's eye, and ensure that each necessary part was there when needed.

Compressed air, for example, would be needed to drive rock-drills. That meant diesel compressors. How many? How big? Where were they to be found? How were they to be shipped? How long would they take to arrive? How much fuel would they burn in a day? What lubricants would they need? Which fuel and oil filters would they need? What substitutes would suffice? What would be the diameter of the air outlet on each? Would it be threaded, or grooved for Victaulic pipe clamps? If the air outlet had a two-inch male pipe thread, a short length of two-inch hose with the correct end fittings would be needed to join this outlet to the four-inch steel Victaulic pipe that would take compressed air underground. This hose would have to be joined to the pipe by means of an adaptor fitting from two-inch male pipe thread to four-inch Victaulic groove. You could not walk into a hardware store and buy such an adaptor. It was a specialty item on two or three weeks' delivery from Ontario. Without that adaptor there would be no drilling, no blasting, men hanging around, bored and irritable, big shots in Vancouver demanding to know why nothing was happening. If John forgot one twenty-dollar hunk of metal that he could hold in his hand, the whole operation might be at a standstill until it or some substitute could be obtained.

That was only one part of one subsystem of many that composed the whole. No detail, however small, could be neglected, forgotten, left to chance, or left to take care of itself. John was the only man in the organization who had the experience to do what had to be done; the crushing weight of this responsibility secretly appalled him. He knew, too, that some others were not fully aware of what they had entered into and that from time to time he would have to carry that weight as well. Nevertheless the challenge of the task fired him with enthusiasm (as did the likelihood of making a substantial sum of money). As the crew assembled, he would come to realize that other competent men would strive with body and soul on his behalf and looked to him only for a leadership that did not flinch from the task at hand.

His visit to the site completed, John made his base in Sudbury, Ontario, for a few days. Canroc's main source of rental machinery was the colossal mining industry of northern Ontario. John had traced a suitable drill jumbo, which was available for rent, to a company that mined barite near Matachewan.

When John worked in the Porcupine, he had from time to time noticed a range of low hills far away to the southeast— a faint corrugation of an otherwise stupefyingly flat horizon. He was told that those were the hills around Matachewan. In among those low hills was a small barite mine that worked steadily from year to year. In all the time he lived in northern Ontario, he never knew of its existence.

Four hours after leaving Sudbury in a rented car, he passed through the village of Matachewan with a feeling that he had crossed the edge of the world, so lost was the settlement in the wilderness. The barite mine was situated a few miles from the village on the banks of the Montreal River, which he crossed on a timber bridge. The mine foreman gave John a friendly greeting and sat him down to a meal in the cookhouse while he put his men to work for the afternoon.

The mine consisted of nothing more than a few ramps and adits and an open cut on a fifteen-foot vein of white barite that striped the brown and gray rock hillside. The crew worked only on day shift. At the mine was a bunkhouse camp that was unused except for a cookhouse where a French Canadian girl cooked the men's lunch; she served John a rib-sticking meal of pea soup, stew, and coffee. The girl spoke no English at all. John's knowledge of French just sufficed for him to explain himself and make complimentary remarks about the cooking. After lunch he inspected the machinery, returning to Sudbury as the warm afternoon turned to a golden evening.

Experience had shown that the rock and bog around the exploration site at Burnt Island Pond was impassable to all but certain specially constructed vehicles. Even these tore up the surface so quickly that their regular use could not be countenanced. Every scrap of equipment had to be assembled at the little town of Port aux Basques and airlifted in by helicopter. When machines weighing eleven tons had to be moved, it was no task for the Hughes 500 used for moving men and supplies. A Super Puma was the most powerful machine readily available in eastern Canada, but its rental rate was over $5,000 an hour. Canroc was therefore allocated two days with this machine to airlift all their heavy equipment in to the site. Consequently all major equipment had to be assembled in advance of this airlift, with no second chance. The scooptrams and drill jumbo would have to be dismantled into assemblies weighing no more than 9,000 pounds, airlifted on slings under the helicopter, and reassembled on the rock dump outside the portal. Reassembly would be no small task in a fully equipped workshop with a ten-ton overhead crane. Out in the open it would be a real test for the ingenious Newfoundland mechanics.

The event had something of the air of an invasion. The people of Newfoundland were astonished at these men who came out of the west and descended on them from the sky. Corner Brook was their ground zero. They demanded, as of yesterday, great quantities of strange equipment and previously unknown supplies and in return were prepared to spend money as if there was no tomorrow. Meanwhile, flatdeck trucks rolled off the ferries at Port aux Basques bearing large, dirty pieces of machinery and unloaded at a staging area outside the town. Derek Mercer, Ed Vardy, Dave Noseworthy, and Sam Kettle were equal to the task and, by working all hours of the day and night, helped John to assemble the mass of equipment at Port aux Basques.

The client company was operating the camp and arranging for the helicopters, and so the day was set. John drove down from Corner Brook, where he was staying in the company's apartment, and lodged in Port aux Basques, where his troops were assembling. Ed Vardy was there, and the two of them spent a convivial evening over a bottle of whisky. They were both tired and nervous, as this was the culmination of a string of twelve- and fourteen-hour days and seven-day weeks. A whole house of cards had been built out of arrangements for certain people to be at certain places at certain times with certain tools and equipment. A degree of nervous anticipation was excusable.

Next morning the assault went in, except that no enemies arose to meet them from the crags and heather of the windy moors. John met the first four men of his crew. They sized each other up and liked what they saw. To them Canroc had been an advertisement in a newspaper placed by an unknown mining contractor—they had worked for most of the known contractors—and a voice on the telephone offering them work at a camp in a remote part of their beloved Newfoundland. To John, they too had been voices on the telephone and applications written for them by their wives, sisters, and mothers. In New-foundland work was for men, writing was for women. For the most part they had answered the call with only the most cursory questions as to how much and when they would be paid, and how they would be housed and fed. They had packed their gear and come. As the crew built up, others would come with just as little fuss and just as few questions. It was John's responsibility to lead them into the wilderness and into the darkness underground.

The south coast of Newfoundland is renowned for its wind, rain, and fog, but on this day the sun shone from a cloudless sky; a cool wind brought with it the smells of seaweed and salt water and the crying of gulls. The scene out on the Isle aux Morts road was one of intense activity, which attracted a fair crowd of spectators. Two helicopters were coming and going. The downwash from the Super Puma's thrashing blades kicked up stones as well as clouds of dust and trash as it struggled to lift the heavy machinery assemblies that had to go in to the site.

Half a dozen men were dismantling machinery, attaching cable slings, and bundling loads in sling nets. John left Ken and Wilf on the road and took Ray and Carl with him to fly to the site in the Hughes 500. Their arrival was singularly lacking in the dignity that should have attended such an occasion.

They landed on a blueberry patch near the portal. John sprang out, tripped in a boghole, and went sprawling and groveling to get away from the slashing rotor blades close overhead. Carl laughed so hard that he hit his head against a stanchion and crept out with more care. Ray stepped into the same boghole and also went sprawling.

Equipment began to rain from the sky as the two helicopters shuttled back and forth. Work went on late into the long evening and all the next day, a Saturday and a Sunday. As fast as equipment was slung in, it was stored here, stacked there, assembled, tested, started up, put into service, or whatever was appropriate to its function. And so they launched the mine at Burnt Island Pond.

23

A FISHY BUSINESS

The decline was soon pumped out, and the necessary facilities were set up. Eight days after they landed on the site, Ken trundled the drill jumbo down into the portal one afternoon; he fired the first blast a few hours later. A series of dull reports smote the summer air, followed by billows of acrid blue smoke that drifted out into the sunlight. The date was July 27.

John and his crew had until the middle of November to drive 2,000 feet of decline and drift—just three and a half months. The pressure to complete the work in that time came from the fact that work would become impossible on the exposed moors from the middle and later part of November on into winter. If winter came fierce and early, the available time would be shorter yet.

The stories of winter weather on the southwest coast of Newfoundland were alarming. Dominating everything was the wind, which had been known to blow freight trains off the track where the railroad ran beside the sea. Wind speeds greater than one hundred miles per hour were far from rare in winter; on one occasion the wind on the hill was measured at eighty knots by reading the airspeed indicator of a parked helicopter. Such a wind could kick up snow to produce an opaque maelstrom near ground level in which visibility was but a few feet; without guide ropes a person could become lost in the blizzard.

The managers of one company decreed from the comfort of their Toronto head office, against the pleas of their field superintendent, that diamond drilling was to go on into December. That was fine until the wind made a tumbleweed out of their drill rig and the crew spent a night crowded into the cookhouse trailer, without heat or light, hoping that their combined weight would prevent it from being rolled over by the wind. They were

rescued by helicopter the next day, abandoning their site to the fury of the winter.

Canroc's client had taken the decision to drive the decline quite early in the year, but they had fumbled and hemmed and hawed until May. That was when a mining crew should have been moving onto the site, but they never gave Canroc the green light until the end of that month. Until that was done, nothing could be arranged, no money spent, no supplies or equipment ordered, no men hired. Starting in their Vancouver office on May 29, Canroc assembled all the necessary equipment and supplies 3,000 miles away in Port aux Basques in seven weeks, in spite of a week's delay when the client company's president suddenly decided to review the whole project and discuss it with his board of directors.

John was the field superintendent; the whole weight of getting the job done in the time available sat fairly and squarely on his shoulders. So frenetic was the pace of the work that the events of a month's normal existence were compressed into a week; what seemed in retrospect to have taken place over a week was shown by the logbook to have happened in a few days. John had had no weekends off and very little rest for more than a month, with none to come in the immediate future. The job went on at a remorseless pace. The camp food was good, plain, and abundant. John devoured it ravenously as fuel to drive his system through days that began soon after dawn and went on long after dusk. Sleep was eight hours of exhausted oblivion, a mere punctuation mark between each successive day. The crews worked twelve-hour shifts around the clock. Machinery was just as likely to break down at night as by day, and the wretched mechanics worked in the wind and rain, plastered with grease, all day and half the night, sleeping when they could.

The client company assigned a young civil engineer to the site as a surveyor and to work as John's "engineering assistant." He and John shared a room in one of the trailers. Bruce was a good fellow, but one day he committed the cardinal sin of telling John that he had not done something because he "did not think it was urgent." The reply that scorched his ears was so sarcastic and so profane that John later heard him babbling about the job in his sleep.

More and more men came to swell the crews, not only John's mining crew, but also men to do diamond drilling and line cutting on the surrounding moors. Tents mushroomed around the corral of trailers. The tents were hot by day and cold by night. By day they let in light and all the noises of the camp, especially the drone of the helicopter, which blew dust all over everything, including food on the men's plates as they sat eating in the cookhouse. By night they rattled in the wind. Brian Warford remarked after one stormy night, "My tent, ol' man, she was a-shiverin' and a-shakin' like a dog shittin' herrin' bones." John had inspected this setup before bringing in his crew. His comments were so corrosive that the relatively solid trailers were at once preempted for the mining crew, and the geologists, diamond drillers, cooks, and bottle washers were unceremoniously bundled out into the tents.

Southwestern Newfoundland was not an easy place in which to carry out an undertaking such as this. Most mining supplies had to come from Ontario, a thousand miles away. In a major mining center, such as Sudbury, any machine, spare part, or consumable item that has anything to do with mining can be rustled up within hours. Even in the Clear Creek country of Colorado, so it is said, hardware stores stock Eimco parts. In southwestern Newfoundland there was practically no mining; few people knew anything about mining supplies and certainly did not carry them in stock. At best some substitute might be available, but most substitutes will not stand up to the rigors of underground service for long. Supplies from the Canadian mainland came on trucks shipped across on the ferries. The ferries docked in Port aux Basques, but some trucks did not unload until they reached Corner Brook, 100 miles away. Anything sent in by the appallingly expensive air freight came through Stephenville, 75 miles away, or Deer Lake, 30 miles the other side of Corner Brook. Besides, if something breaks down, spare parts are needed at once or sooner, and ten to fourteen days is a long time to wait. By that time something else will have broken down; if machines break down faster than they can be repaired, it does not make for fast or efficient progress.

Sam Kettle, who lived in Port aux Basques, was the local expediter, shipping agent, and universal fixer for the project. The devotion that he and his wife, Lydia, lavished on the job and the people connected with it knew no bounds. If a truckload of equipment had to be met as it came off a ferry at 2:00 A.M., Sam would be there to meet it. If something had to be fetched from Corner Brook, Sam fetched it. If someone had to be brought to the helicopter pickup point, Sam brought them. If someone had to mount guard over equipment overnight before it could be airlifted in, Sam did that too. If someone needed a bed for the night when both hotels in town were full, they found it under Sam's roof. If Sam could not handle something alone, he enlisted the help of his numerous friends and relations. John said that Sam slept like a snake, with one eye at a time.

Somehow the necessary messages were passed through the grapevine and whatever was needed was done. Time after time, John would hear some voice, known or unknown, on the radio telephone, "John? John? That you, boy? Look, Ed Vardy, 'e sez that thing you was lookin' for, it'll be on the Air Canada into Stephenville this evenin'. So I got 'old of Eddie Jesso an' 'e sez 'e'll bring it with 'im when 'e comes in tomorrow. OK, boy?"

Even Sam could not defeat Murphy. Nothing that could break down did not do so. Several things that could not go wrong in any normal circumstances did so now. Factory-new electrical gear shorted out on first use because of internal defects in the wiring. Factory-new jacklegs refused to function. The vitally needed spare part was in the one parcel that Air Canada's freight department managed to lose. It was the bundle of fourteen-foot drill steels on special order from Ontario that fell out of a helicopter sling into the bog and was never seen again. By the operation of Murphy's Laws everything broke down on Friday afternoons, just as the outside world was shutting up shop and going home for the weekend, leaving John and his crew to struggle on as well as they could.

One morning the night shift crew greeted John with the news that one of the scooptrams had come apart into two pieces down the decline, with its bucket full of rock. As a loaded scooptram weighed fifteen tons, the pieces would weigh about seven and a half tons each. It would be a neat little problem how to get the two pieces back together again underground, get the machine out to the surface, and effect whatever repairs might be needed. John wolfed his breakfast and headed for the portal to view this disaster. Carl Loveman remarked airily, "Oh, that's happened before. I seen a ST-5 do that at Sussex, New Brunswick." John rounded on him in exasperation. "Yes, but you tell me why everything that ever went wrong in any mine in Canada has to go wrong in this @#$%ˆ& place, here, now."

Nor could Sam or any of them influence the weather. When a southeast wind blew over the cold Labrador current, fog blanketed the south coast of the island. One electrician flew in to a remote site by helicopter for a two-hour job; he stayed there for seven days. Most helicopters used in the bush are equipped for day visual flying only. The rental rate on a helicopter equipped for instrument flight in fog or darkness is astronomical. They knew that because they had one lined up in St. John's for emergency medical evacuation with the camp's coordinates programmed into its Loran-C.

Even though that summer was the driest in forty years, the southeast wind still blew at times and brought with it fog, which covered the coast for days on end. Fog would come rolling up the Isle aux Morts River valley, and soon the camp would be buried in gray vapor. Although a helicopter can creep along close to the ground in fog, it is too slow, too dangerous, and too exhausting for the pilot to contemplate except in dire emergency. Besides, it was easy to become lost over the featureless moors, even in good visibility.

The helicopter pilots attached to the camp for their one-month stints worked desperately long hours, often under the most trying circumstances. John said that you could spot the helicopter pilot by the spaced-out look on his face. So, when fog rolled in, the exhausted pilot would heave a sigh of relief and flop onto his bed. Sometimes he was lucky enough to be caught by fog at Port aux Basques, which would give him a brief respite from the life of the camp. Because of the fog they needed to keep a ten-day supply of food, diesel fuel, and explosives on site. As long as these commodities were on hand, driving of the decline could go on.

Sometimes the weather had other tricks up its sleeve. Watching the windy sky of Newfoundland was like watching a time-lapse movie. Weather changed by the minute; clouds and fog came and went; torrential rains fell and then ceased as suddenly as they had begun. John had just awoken one morning, when his disbelieving ears caught the rumble of thunder. Cursing into his pillow, he knew that this particular storm—the only thunderstorm that he heard or saw all year—was bearing down on him personally and would do him or his crew some harm. He was right. The storm passed over the camp; lightning struck like gunfire, and it went its way, passing directly over the portal. When John telephoned his daily report to Vancouver later that morning, he remarked, "Both scooptrams and the jumbo are broken down, and the mechanic got hit by lightning, but otherwise everything's fine."

Les Patey had been kneeling on top of a scooptram outside the portal when a bolt of lightning hit it. As he had not been touching the ground, he was affected more by shock and concussion than by electrocution. After depositing his breakfast on the ground, Les insisted on working for the rest of the day. The equipment situation was too desperate for John to object, but he conferred with Brian Warford, the first aid man, and that evening they shipped Les out under lurid protest to spend the night under surveillance in hospital at Port aux Basques. When he returned the next day, his description of the nurses had the crew's rapt attention.

The client company was exploring large tracts of the moorland surrounding the camp. John's mining crew was merely an addition to the geologists, line cutters, diamond drillers, geophysics operators, helicopter pilot, cooks, and people needed to hump supplies, run the camp, repair breakdowns, and maintain some semblance of cleanliness.

The original camp had been intended for eighteen to twenty men at most. As late as May it had been nothing but a collection of damp, moldering wreckage, smelling of decay, through which wind blew and rainwater seeped unhindered. Within a matter of weeks it was supposed to house first twenty, then thirty, then forty men, all of whom were expected to put in long hours at hard, dirty work for continuous periods of two and three weeks at a time. The camp was supposed to provide the means for them to eat, sleep, and wash themselves and their clothing in some degree of shelter and cleanliness, but it offered meager respite from mud, dust, rain, wind, and noise. No less than six separate companies were represented in camp, all contracting and subcontracting to each other. The surprising thing was that it all worked. The Newfoundlanders were used to working hard and living rough. To a man used to fish boats or to sleeping out on the moors under a tarpaulin, the camp was more than adequate. To a man used to a steady diet of fish and potatoes, camp food bordered on the exotic. If a man could just earn some money without leaving Newfoundland, he asked for little more. And if the roof leaked, who cared?

The client company's vice-president of exploration was in camp one day on a visit from Vancouver. John was in the cook trailer at lunchtime, packing away a solid meal of stew. The room was redolent of the steamy smells of food and wet clothing. The trailer shook as gusts of wind buffeted it and deluged it with rain. Walking through the trailer, the veep received a dollop of water dead center on the crown of his head. Turning to John, he asked the asinine question, "Does the roof leak?" John had never been greatly inhibited by respect for rank or title, and the opportunity was too good to miss. He grinned expansively. "Hell, no!" he replied, "Only when it rains!"

Laughter was a sure antidote to the fatigue and discomfort of life out there on the bog. Countless little episodes day by day had the participants laughing until their stomach muscles ached and tears streamed down their faces, and until they crept away, whimpering, because it hurt too much to laugh any more. Laughter, and at times anger, took on extravagant dimensions.

The client company participated in driving the decline both by running the camp and arranging helicopters and by providing specialist manpower if required. Partly they reckoned that John's task was beyond the power of one man to sustain; partly they felt they should have their own emissary to keep an eye on how things were going. They sent Bud McKenna. Bud was in his early sixties; he had sunk shafts and driven drifts and declines and raises all over Canada and had worked for every contractor in the business. Bud arrived with the intention of helping where help was needed and keeping his mouth shut where it was not. He and John became immediate friends. Bud remarked that, after he and John had known each other for a day, it seemed as though they had been working together for years.

John made a habit of going down the decline at ten o'clock every evening to visit the night shift crew before going to bed. One night the weather was foul, blowing a gale and pelting with rain as he made his way down the road from the camp. The crew were at the bottom of the decline, drilling on the face. Glen, the jumbo driller, enthroned on the back of his steel dinosaur, stood in watchful contemplation, silhouetted against the fog and oil mist that churned from the thundering machines and swirled about him. He turned at John's arrival; his gold tooth glinted beneath his handlebar mustache. Frank, the crew leader, leaned against the wall in that same watchful contemplation. After talking to them briefly, John walked back up the decline and out into the rainy darkness, sloshing and slithering up the muddy road toward the lights of the camp. The gale pummeled him as it roared over the black-dark moors; rain rattled on his oilskins like flying grit.

He poked his face in through the door of the room that Bud

shared with Gordie Denison, one of the mechanics. Both were lying on their beds. Gordie lowered his newspaper and peered at John over his glasses. Bud was in the classic bunkhouse posture: supine, hands clasped behind his head, staring at the ceiling. He raised his head. "Hullo, John! Come on in." John fought with the wind for control of the door and shut it behind him. Picking a piece of newspaper off the gritty floor, he laid it on the chair with exaggerated care, sat down, and began to drip rainwater. The room was warm and snug because Bud and Gordie had stuffed the cracks with newspaper and had scrounged an extra heater that added its load to the camp's overloaded power circuits.

They could hear each wind gust as it came roaring over the bog to hit the trailer with a blast as if an express train had gone by. Rainwater bubbled in around the window frame. They talked about the job for a while, then fell to yarning about the strange places and stranger people they had known. Gordie asked in curiosity, "John, where are you from?" Having lived in so many different places, John had to consider this question for a while. Bud chuckled. "He doesn't know. He's a tramp miner like the rest of us." John laughed. It was true, and in a way, coming from a man like Bud McKenna, the words were a benediction.

Bud McKenna was there for another reason. John knew it; Bud did not. In the very beginning Canroc believed that their client was the company that owned the property, a company named, shall we say, Fish Creek Explorations. They discovered, however, that Fish Creek Explorations was owned by a larger company, Pisces Resources, which was based in Vancouver. Canroc's contract was with Pisces. Pisces took Canroc's invoices, marked them up by a substantial percentage as a "management fee," and passed them on to Fish Creek Explorations.

The president of Pisces was a friend of the owner of a mining contracting firm named Lone Star Contracting also based in Vancouver. Having worked in British Columbia for five years, John knew that Lone Star never went up the front stairs if the back stairs would do equally well. As they were a successful contractor, this was obviously an effective way of doing business.

At discreet little lunches in select watering holes in downtown Vancouver, the presidents of Pisces Resources and Lone Star Contracting agreed over wine and white napkins to take the job at Burnt Island Pond from Canroc and give it to Lone

Star. Canroc would be given a fair trial and shot; Lone Star would not have to incur the expense of bidding on the work or go through the toil of getting it started. Just as the job was beginning to move along, as most jobs do after their initial troubles, Lone Star would walk in, "get this mess cleaned up," and take the credit for its successful completion.

The mine at Burnt Island Pond was also a pawn in a power struggle within the senior management of Pisces Resources. The vice-president of exploration was in charge at first and awarded the work to Canroc. The vice-president of mining successfully ousted him and, besides, was a friend of a vice-president of Lone Star Contracting.

Bud McKenna was an employee of Lone Star Contracting. John knew who signed Bud's paychecks before Bud ever appeared on the scene and therefore knew why he was there. John knew that Bud was too upright a man to realize what was going on and never let it taint the friendship that sprang up between them. As soon as John became aware of Lone Star's presence in the background, however, he realized intuitively what would happen and warned Clarence back in Vancouver. Clarence could do nothing about it, but the picture became clearer as the high rollers with Pisces exerted gradually increasing pressure on Clarence to abandon the contract.

Six weeks after landing on the site, Canroc had set up their equipment, reopened the existing portal, and driven 500 feet of new 10' x 10' decline on a 15 percent grade down into the rock. They had predicted that they would advance the decline by eighteen feet every twenty-four hours. They had done this by analyzing the maximum performance of their equipment and taking 60 percent of that to allow for the normal operation of Murphy's Laws. John knew from experience that Murphy generally had a success rate of about 40 percent, a rate that decreased in proportion with the experience of the crews, the quality of equipment, and general operating conditions at the site. This time, however, Murphy had outdone himself and they averaged only fourteen feet a day, which left them seven days behind schedule.

Any job has a "shakedown period"—often longer than a month—in which the crews learn how the rock behaves, equipment is fixed up so that it works in the desired manner, and people learn how to work with one another. One of John's men remarked in the cookhouse one day that he had recently been offered a job at an operating mine but had refused it. The fact was that Bud and John and the crew liked each other; Canroc was already well thought of in Newfoundland, and between them, they had stuck together and toughed it out. After a month the rate of advance was climbing steadily to sixteen and then twenty-two feet a day.

The routine monthly visit by the big shots from Vancouver came around. Two veeps and one or two lesser luminaries from Pisces Resources came along. Naturally, as Bud was a Lone Star employee, a Lone Star veep traveled 3,000 miles to inquire as to his well-being.

In the camp's own half-humorous, half-cynical slang, an employee was termed a unit; a big shot was known as a wheel. "Hey, there's another Fish Creek unit coming in tonight," yelled hurriedly at someone's departing back, would mean all things to all men. The helicopter pilot would know that someone had to be collected from Port aux Basques that evening. The cook would know that another live body had to be kept that way. Dan the Man, ready to weep with frustration, would disappear, leaving vile oaths hanging on the air, to rustle up a vacant bed and some bedding under some sort of shelter in his overstretched camp. The Fish Creek chief geologist would scratch his head and wonder who had been sent to him and why. (No one ever told him anything.) The camp at large would know that a new face was about to appear who would soon cease to be a faceless "unit" and would make his own contribution to their community.

Wheels were different. As one of the diamond drillers remarked over his plate of fishermen's brews, "Like seagulls. Fly in, shit on everybody, fly away again."

This time the wheels were delayed in Port aux Basques by a day of thick fog. The fog lifted late the next morning to become a gray overcast raining steadily. The wheels were airlifted in around lunchtime. The units gathered around one end of the cookhouse, picking their teeth, to watch the helicopter land. The wheels climbed out into the mud. Briefly the two groups surveyed each other, the units with their weatherbeaten faces, their faded bush clothing hanging loosely about them, the wheels looking as if they had stepped out of the show window of a men's leisure wear store—and were unsure of their wisdom in having done so.

The wheels went beetling all over the property, down the decline, prodding and poking among the machinery. John tagged along. He knew this procedure from long experience. If the superintendent's existence was acknowledged at all ("Here, hold this"), he was generally treated with scant courtesy. Wheels liked to impress each other by arguing over some question that the foreman or the miner doing the job could answer in a few words. Sometimes a wheel would browbeat the foreman or superintendent just to show the others how important he was. After the wheels had gone to the cookhouse, or perhaps to make like a seagull all over the chief geologist, John tracked Bud to where he was lying on his bunk, looking at the ceiling.

"Well, Bud, what do you think?"

"I think those bastards are up to something," Bud replied, without removing his gaze from the ceiling.

They were. John discovered later that one of the purposes of the visit had been to hand the job over to Lone Star Contracting. John's little fiefdom was to be taken from him and given to someone else. The crew received the news with varying degrees of anger or resignation. Paradoxically, those whom John had blasted most savagely with his wrath for some neglect or misdemeanor were most put out by this turn of events. Word gets around a place like western Newfoundland very quickly, and Canroc had made for themselves a reputation for square dealing that was known all over that part of the island. But when the wheels made a deal, the units were not asked for their opinion.

John knew that Bud would not stay for long; he had served his purpose. A Lone Star unit would come in and kick John's men around, fire some, and replace them with his own cronies. So it turned out.

The camp was supposed to be "dry"—no alcohol—and the arrangement worked well. The rule was well accepted, and as far as John knew, it was never broken, nor were there many complaints. His crews worked a rotation of fourteen days on and seven days off, although most of them came from northern Newfoundland and spent a day traveling each way. One crew of three men would head for the liquor store in Port aux Basques as soon as the helicopter flew them out on the first morning of their leave. One man would volunteer to drive while the other two concentrated on getting seriously drunk—"drunk by Corner Brook, dead drunk by Springdale." But as far as John knew, they never drank in camp.

Lone Star Contracting seemed to specialize in employing

drunks. Their superintendents were the drinking, fighting type and cared nothing for the wishes of those running the camp. The Lone Star superintendent himself brought drink into the camp, so it was said. Some of John's men were fired without a word of thanks for what they had done. A whole shift crew quit all at once.

The decline went on, but at far greater cost to the client. Canroc had tried to do the job economically for their client; it turned out that the client could not care less. Pisces took the larger invoices from Lone Star, marked them up as before, and passed them on to Fish Creek Explorations. In fact, the bigger the invoices, the more money went to Pisces. Bigger machinery was flown in by Super Puma in an airlift that had been denied to Canroc. Alleged "safety deficiencies" that had been used to bludgeon Clarence were suddenly forgotten.

The people of western Newfoundland knew a dirty deal when they saw one. The men outside, who had put so much effort into supporting the job, were no longer so eager and looked to their own interests. The high rollers in Vancouver at once complained that Newfoundland people were idle and dishonest. The wheels at Fish Creek and Pisces fought among themselves. Within a year both Lone Star and Pisces were owned by others.

John departed as soon as Lone Star moved in. He showed the incoming foreman around the place, as requested, but the man had the build of a locomotive and brains to match, so John might as well have talked to the wall. Rather than waste his time, John said good-bye to his men, thanked them for all their hard work, and returned to Vancouver.

The morning of his departure was cool and bright. It resembled another morning, a long time ago, in another land of miners and fishermen 2,000 miles away across the Atlantic, where the salt wind blew wild and strong from the sea, and surf rumbled and rustled against the rocks as it did against the gray crags of Newfoundland. John sat in the cookhouse, drinking coffee, while the helicopter flew diamond drillers out to their rigs before taking him to Port aux Basques. The cookhouse was deserted, apart from the laborers who worked around the camp. They had bidden him farewell and had shaken him by the hand, but now the farewells had been said, and they moved to and fro in the course of their work as if he were not there. He was no longer a part of their lives; nothing more remained to be said. So he sat drinking coffee from a thick china mug while the plastic sheeting over the window behind him flapped and slatted in the wind.

A swelling drone trespassed on John's hearing. The helicopter was returning from the drill sites on Windowglass Hill; the sound of its rotor broke over the camp like a surf as the pilot came in to land. Whenever he heard the distinctive sharp drone of a Hughes 500 in years to come, John's mind would instantly take him back to the mine at Burnt Island Pond. A few minutes later the pilot beckoned to him from the doorway. Leaving the empty mug standing by itself on the table, he went out into the sunlight and the wind, where cloud shadows raced over the brown moors, just as they had on that bright morning in west Cornwall so long ago.

EPILOGUE

On the following day John left Newfoundland to return to Vancouver. After its clamorous departure from Deer Lake the airliner climbed six miles into the sky through floors of gray cloud until it hung in space, floating above a white plain that unrolled beneath its outspread wings. John sat back, running his eye over the dully gleaming wing, the wisps of cloud above, and the rolling vapor beneath, daydreaming.

Once upon a time he might have been shoehorned into an office job in London, capital of England—a grimy, sprawling metropolis of 12 million people. If he had been crimped, willingly, unwittingly, or by force of circumstance into a job in a city office, he would have been no happier than a caged animal.

Considering the past fifteen years, he could look back on a rich abundance of fun, adventure, and laughter. He thought of the people whose friendship he treasured, whom he would never have met but for the mining industry. He remembered the floodlit halls under the Rammelsberg, the aurora borealis over Yellowknife, the gold in the rock of the Canadian Shield, and the snow on the Sangre de Cristo Mountains under the glare of a Colorado moon. He reflected on the good friends and good fortune that had been his ever since he met his vocation face to face that first shift at South Crofty. He looked back on all these things with contentment, envying no one and regretting nothing. The aircraft flew on above the clouds, the brown moors of Newfoundland invisible below.

The clouds beneath the cruising aircraft began to part into tufts and streamers, each one floating by itself in the void. The brown land, veined with rivers and spotted with steel-gray ponds, came into view far below. Down there was the coast.

John sat up. There was Port aux Basques, identifiable in every detail—the motel where they landed with the helicopter, the yard where they had stored their equipment before the airlift; he could even see where Sam Kettle's house was.

John pressed his face to the window. His eyes followed the Isle aux Morts River up from the sea, its valley marked by trees growing in its shelter. There was the bend in the river. There, in the crook of the bend, was what his eye sought, a tiny, pale speck on the uttermost limit of vision. That speck was the camp. Down there in that bland, brown landscape was the job, even now fading behind the mists that were once more gliding beneath the aircraft. Down there was the decline, the mud and the darkness, the smells of dynamite and diesel smoke, the scooptrams growling and bellowing on the ramp, the pealing thunder of the machines, and Wilf, Frank, and Bernard, and the boys punching the rounds into the hard rock of Newfoundland.